The Ethics of Information Warfare

Law, Governance and Technology Series

VOLUME 14

For further volumes:
http://www.springer.com/series/8808

Luciano Floridi • Mariarosaria Taddeo
Editors

The Ethics of Information Warfare

Springer

Editors
Luciano Floridi
Professor of Philosophy and
Ethics of Information
Oxford Internet Institute,
University of Oxford
Oxford
United Kingdom

Mariarosaria Taddeo
Politics and International Studies
Research Fellow in Cyber Security and
Ethics, PAIS, University of Warwick
Coventry
United Kingdom

ISSN 2352-1902 ISSN 2352-1910 (electronic)
ISBN 978-3-319-38006-3 ISBN 978-3-319-04135-3 (eBook)
DOI 10.1007/978-3-319-04135-3
Springer Cham Heidelberg New York Dordrecht London

Printed on acid-free paper

Springer is part of Springer Science+Business Media (www.springer.com)

The Ethics of Information Warfare—An Overview

"By the word 'information' we denote all the knowledge which we have of the enemy and his country; therefore, in fact, the foundation of all our ideas and actions [in war]."
(C.Von Clausewitz, F. N Maude et al. 2008, p. 81).

This volume collects twelve original contributions addressing some of the most important ethical problems raised by Information Warfare (IW), the complex set of new phenomena associated with the use of Information and Communications Technologies (ICTs) in fighting scenarios. IW is redefining how war is waged. In doing so, it is reshaping the concept of war itself, raising new ethical problems and challenging old solutions. These transformations are at the core of the current debates in research fields such as ethics, philosophy of technologies, war studies, and political philosophy. The main purpose of this volume is to provide an interdisciplinary investigation of some of the most compelling ethical problems posed by IW and to present innovative analyses for their solutions.

Before the pervasive dissemination of ICTs, the expression 'information warfare' referred to the importance of information, understood as the semantic content (Floridi 2010), within military strategies. Information as semantic content is relevant to war-waging both in relation to intelligence-gathering and as a means for propaganda aimed at demoralising the enemy's military forces and civilians. However, with the advent of the information revolution and the capillary dissemination of ICTs, the role of information in warfare radically evolved. ICTs further support war-waging in two new ways: by providing unmanned weapons to be deployed on the battlefield—like drones and semi-autonomous robots used to hit ground targets, defuse bombs, and patrolling actions—and by creating an entirely new battlefield, called the 'cyber domain', where warfare is waged with software tools, e.g. computer viruses or security packages. During the past two decades, such new uses of ICTs in warfare proved to be convenient and effective and gained a central role in militaries strategies. Nowadays, IW indicates a heterogeneous phenomenon concerning the deployment of robotic weapons, of cyber weapons, and the use of ICTs to foster coordination among militaries on the battlefield and for propaganda, the so-called C4ISR (integrated command, control, communications, computers, intelligence, surveillance, and reconnaissance) (Libicki 1996; Taddeo 2012).

The rise of IW is not surprising. Historically, technological breakthroughs determine changes affecting the structure of both civil society and military organisations. As described by (Toffler and Toffler 1997), this was the case with the Neolithic revolution, when human beings first made weapons out of wood and rocks, and with the Industrial revolution, which provided the means for industrialised warfare and for the dissemination of weapons of mass destruction. The Information revolution is the latest example. It has changed our activities in several ways and at several levels (Floridi 2010). The use of ICTs changed the way individuals manage their communications and daily practices, from working and reading books, listening to music and driving. At a social level, ICTs reshaped social interactions; at the institutional level they provide new tools for the management of information and bureaucracy (Ciborra 2005; Saxena 2005); and when considered with respect to warfare, ICTs determine the latest revolution in military affairs. In this sense, IW is the warfare of the information age.

Nonetheless, it would be misleading to consider this new type of warfare simply as the latest evolution of war fighting techniques. For IW engenders radical changes, which concern the very way in which we understand war, not just how it is waged. War is traditionally understood as the use of violence by a state through the latter's deployment of military forces, in order to determine the conditions of governance over a determined territory (Gelven 1994). As Oppenheim put it: "war is a contention between two or more states through their armed forces, for the purpose of overpowering each other and imposing such conditions of peace as the victor pleases", (Lauterpacht 1952, p. 202). The choice to undertake a (traditional) war usually involves a substantial commitment, given its heavy human, moral, economical, and political costs. Such aspects of war-waging have been radically changed by IW, which provides the means to carry out war in a completely different manner. The changes determined by IW are of astounding importance as they concern both the way the military and politicians consider and wage war, and the way war is perceived by the civil society. Like traditional warfare, IW is very powerful and potentially highly disruptive. However, unlike traditional warfare, IW is potentially bloodless, cost effective, and does not require military expertise. In short, ICTs have modified the costs of war, and hence our understanding and evaluation of them.

Furthermore, the information revolution brought to the fore a new domain, the cyber domain, which has become an important part of the environment in which we live and interact and plays crucial role for the development of contemporary societies (Floridi 2010). The management of national health systems, the regulation of energy, water and food supply-chains are only some examples of the aspects of contemporary societies that largely depend on the efficient functioning of the cyber domain. In this context, the ability to control, disrupt or manipulate the enemy's informational infrastructures has become as decisive, with respect to the outcome of conflicts, as weapon superiority. Ethical analyses of IW need to take into account these aspects, for they pose important ethical problems concerning, for example, the moral stance of the entities existing in the cyber domain and the moral responsibilities for the actions performed by autonomous artificial agents, such as cyber viruses or robotic weapons, and increasingly hybrid agents, represented by human-machine systems.

Two interwoven sets of problems are of particular relevance, when considering the ethical implications of IW. The first one concerns the definition of IW and its properties. As Orend puts it in his chapter, there is a conceptual fog shrouding this type of warfare. Scholars are still debating on issues such as the nature of non-kinetic cyber attacks (Schmitt 2008) (Arquilla 1998), the definition of IW, its long-term effects on the concept of war (see, for example, Dipert's chapter) and its role in the future development of international politics and economy. Casting some light through this fog is the preliminary and necessary step toward the solution of the second set of problems. These are ethical problems that range from the consideration of the most adequate ethical framework to prescribe principles for conducting a *just* war (Dipert 2011) to the solution of more applied issues. In this respect, three categories of applied ethical problems are at the centre of the contemporary debate on IW, attracting the attention of both ethicists and policy-makers; these are the *risks*, *rights* and *responsibilities*—the 3R problems (Arquilla and Ronfeldt 1997; Taddeo 2012).

Risks. The risks involved in IW concern the potential increase in the number of conflicts and casualties. ICTs-based conflicts may be virtually bloodless for those involved. This advantage has the drawback of making war less problematic for the force that can implement these technologies, therefore making it easier not only for governments, but also for criminal or terrorist organisations, to engage in such conflicts around the world (Arquilla and Borer 2007; Steinhoff 2007; Brenner 2008).

Rights. IW is pervasive since not only can it target civilian infrastructures, it can also be launched through civilian computers and websites. This may initiate a policy of higher levels of control, enforced by governments in order to detect and defend their citizens from possible hidden forms of attacks. In this circumstance, the ethical rights of individual liberty, privacy and anonymity may come under sharp, devaluating pressure (Arquilla 1999; Denning 1999).

Responsibilities. The problem concerns the assessment of responsibilities when using semi-autonomous robotic weapons and cyber viruses. In the case of robotic weapons, it is becoming increasingly unclear who, or what, is accountable and responsible for the actions performed by complex, hybrid, man-machine systems on the battlefield (Matthias 2004; Sparrow 2007). The assessment of responsibility becomes an even more pressing issue in the case of cyber attacks, as it is potentially impossible to trace back the author of such attacks (Denning 2007).

The twelve chapters of this volume address the changes and the problems caused by IW, with different focuses and approaches. The volume is divided into three parts. The first part focuses on issues pertaining to the concept of IW and the clarifications that need to be made in order to address its ethical implications. It includes four chapters: *Fog in the Fifth Dimension: The Ethics of Cyber-war*, by Brian Orend; *The Future Impact of a Long Period of Limited Cyber warfare on the Ethics of Warfare*, by Randall Dipert; *Is Warfare the Right Frame for the Cyber Debate?*, by Patrick Lin, Fritz Allhoff, and Keith Abney; and *Technology, Information, and Modern Warfare: Challenges and Prospects in the 21st Century*, by Wayne McCormack and Deen Chatterjee.

Orend's chapter opens the volume by first addressing the conceptual confusion surrounding IW and outlining some useful clarifications and distinctions. The focus is then shifted on to the analysis of Just War Theory and on how it can be embraced to provide some guidance in waging IW. The contribution stresses that Just War Theory remains the core conceptual framework for evaluating the ethics of political violence in general, and the ethics of IW in particular, but at the same time the chapter acknowledges the patches of darkness and confusion that remain unaddressed by Just War Theory.

Dipert's analysis adopts quite a different approach from Orend's. The chapter focuses mainly on cyber attacks and cyber warfare. He first provides a detailed taxonomy of the different instances of this phenomenon, then discusses possible alternative defensive strategies that may be put in place in the long term by governments in order to guarantee cyber defence.

The contribution by Lin, Allhoff, and Abney concerns above all cyber attacks. They highlight the relation between the policy vacuum concerning the launching of such attacks and the absence of ethical principles that provide guidance for the waging of cyber warfare. The chapter concludes by suggesting that the ethical problems posed by the occurrences of cyber attacks are overcome if such attacks are considered as attacks to privates rather than instances of warfare, which may give rise to private defence, i.e. self-defence by private parties, especially commercial companies, as distinct from a nation-state's right to self-defence.

The chapter by McCormack and Chatterjee addresses the normative and legal challenges that ICTs pose for modern warfare. They first examine the ethical and legal implications of IW in relation to the internal affairs of nations facing armed uprising or undergoing similar violent turmoil. Then, they focus on the ethical consequences of the growing reliance on ICTs in modern warfare and analyse the blurring of the distinction between pre-emption and prevention in self-defence wars.

The second part of the volume collects four contributions focusing on Just War Theory and its application to the case of IW: *Permissible Preventive Cyberwar: Restricting Cyber Conflict to Justified Military Targets*, by George R. Lucas; *Moral Cyber Weapons: The Duty to Employ Cyber Attacks*, by Dorothy Denning and Bradley J. Strawser; *The Ethics of Cyber attack*, by Steven Lee; and *Just Information Warfare*, by Mariarosaria Taddeo.

Lucas' chapter investigates the conditions for preventive cyber warfare. The chapter first distinguishes permissible from impermissible forms of cyber conflicts as well as genuine warfare from criminal or terrorist enterprises. It then stresses the lack of discrimination often encountered in the formulation of cyber strategy and development of cyber weapons. The chapter concludes by considering the case of Stuxnet and arguing in favour of establishing international governance and guidance that (with respect to proportionality, discrimination, and the principle of last resort) may provide regulations for the use of cyber weapons. Lucas argues that cyber warfare is permissible "if it aims primarily at harming military infrastructure, degrades an adversary's ability to undertake highly destructive offensive operations, harms no civilians and/or destroys little or no civilian infrastructure in the process, and is waged as a last resort in the sense that all reasonable alternatives short of attack have been attempted to no avail, and further delay would only make the situation worse."

The contribution by Denning and Strawser also concerns the ethical principles for the deployment of cyber weapons. In particular, this contribution focuses on the international law of armed conflict. It defends the thesis that, at least in some circumstances, the use of cyber weapons not only respects the principles of Just War Theory, but that a "positive duty to employ" such weapons may arise in certain contexts. It is argued that in some cases the option of using cyber weapons is not just permissible for a state, it is actually a moral duty. The moral obligation rests on the consideration that non-kinetic cyber attacks may reduce the risk of bloodshed.

Lee's analysis focuses on cyber attacks and the suitability of Just War Theory for providing some guidance to them. The chapter first investigates the nature of cyber attacks and cyberwar. It then considers cyber attacks on the basis of the principles of *jus ad bellum* and *jus in bello*. Finally, it concludes that while cyber attacks are a novel form of conflict, their ethical dimensions can be understood for the most part in terms of the traditional categories of Just War Theory. At the same time, Lee maintains that the principle of last resort cannot be applied in case of cyber attacks, since each side's fear that the other is about to attack will make it impossible for either side to explore effectively options short of war for resolving the conflict.

Taddeo's chapter has the twofold goal of filling the theoretical vacuum surrounding IW and of grounding the definition of new ethical principles for this phenomenon. The chapter argues that Just War Theory is a necessary but insufficient instrument for evaluating the ethical implications of IW and that a suitable ethical analysis of this kind of warfare may be developed by merging Just War Theory with Information Ethics. The initial part of the chapter describes IW and its main features, and highlights the problems that arise when Just War Theory is endorsed as a means of addressing ethical problems caused by IW. The final part introduces the main aspects of Information Ethics and defines three principles for a Just IW.

The third section comprises three chapters that adopt alternative approaches to Just War Theory for analysing the ethical implications of IW: "*The Ethics of Cyber Attacks*, by Thomas W. Simpson; *Virtue in Cyber Conflict*, by Don Howard; *Armed Robots and Military Virtue*", by Shannon Vallor; and *Deception and Virtue in Robotic and Cyber Warfare,* by John Sullins.

Simpson's chapter addresses the circumstances under which it is permissible to attack ICTs' infrastructures. It analyses Just War Theory in relation to IW and it is argued that Just War Theory is appropriate for assessing the permissibility of cyber attacks in some, but not in all, contexts. The thesis defended is that the concept of *harm to property* provides the right framework to evaluate a great proportion of the moral significance of cyber attacks, which otherwise escapes the principles of Just War Theory.

Howard's analysis adopts virtue ethics to investigate the ethical issues raised by the deployment of tele-operated robotic weapons. The analysis first describes the role of virtue ethics in decision-making processes in general and in war-related circumstances in particular. It then addresses two fundamental questions: whether the technologizing of war made honour and courage irrelevant; and how relying upon the integrity of the cyber warrior (the soldier who remotely controls robotic weapons) may ensure ethical action in cyber conflicts. The analysis concludes by considering how the principles of virtue ethics should be included in the training and evaluation processes of cyber warriors.

Vallor's contribution examines the impact of the progress of military robotics on the perception of virtues in military contexts. While early reflections on the ethical implications of military robotics have focused primarily on utilitarian or deontological considerations, this chapter stresses the importance of an intensive and rigorous treatment of the virtues in the context of military robotics. Three aspects are analysed in detail: the effects of the developments in robotics on the contexts of military action in which moral excellence is displayed; the possibilities and the modes in which robots could embody or emulate virtues, especially prudence and excellence; and the redefinition of the way in which scientists and engineers, both military and civilian, understand their ethical roles in society when 'engineering virtue' in military robotics.

Sullin's chapter considers how robotic and cyber weaponry could be deployed in such a way that our commitments to just and legal warfare are enhanced and not degraded. In particular, the chapter explores the possibilities of designing and developing information technologies that can help us make better decisions on the battlefield. A central aspect of the proposed analysis concerns the concept of deception and the implementation of deceptive strategies by virtuous artificial agents, which are considered trustworthy agents by their 'fellow soldiers'. The chapter concludes by redefining the concepts of deception and trust in relation to artificial agents.

Finally, an afterword by Neelie Kroes concludes the volume. The contribution describes the interests and commitments of the European Digital Agenda with respect to the research for the development of robots to be deployed in several circumstances, of which warfare is one. It also illustrates the goals, namely, ease of use, safety, and autonomy, of the research developed within the European Community and devoted to the design of robots in general, and robotic weapons in particular. The contribution concludes by considering the ethical problems that arise from developing and deploying machines that are autonomous to some degree, such as, for example, the assessment of the responsibility for actions performed by robots.

References

Arquilla, J. 1998. Can information warfare ever be just?. *Ethics and Information Technology* 1:203–212.

Arquilla, J. 1999. Ethics and information warfare. In *Strategic appraisal: the changing role of information in warfare,* eds. Z. Khalilzad, J. White, and A. Marsall, 379–401. Santa Monica: Rand Corporation.

Arquilla, J., and D. A. Borer. Eds. 2007. *Information strategy and warfare: A guide to theory and practice* (Contemporary security studies). New York: Routledge.

Arquilla, J., and D. Ronfeldt. 1997. *In Athena's camp: Preparing for conflict in the information age.* Santa Monica: RAND Corporation.

Brenner, S. W. 2008. *Cyberthreats*. New York: Oxford University Press.

Clausewitz, C. Von, F. N Maude, et al. 2008. *On war*. Radford: Wilder.

Ciborra, C. 2005. Interpreting e-government and development: Efficiency, transparency or governance at a distance?. *Information Technology & People* 18:260–279.
Denning, D. 1999. *Information warfare and security*. Boston: Addison-Wesley.
Denning, D. 2007. The ethics of cyber conflict. In *Information and computer ethics,* eds. K. E. Himma and H. T. Tavani, 407–428. Hoboken: Wiley.
Dipert, R. R. 2011. The ethics of cyberwarfare. *Journal of Military Ethics* 9:384–410.
Floridi, L. 2010. The digital revolution as the fourth revolution. *Invited contribution to the BBC online program Digital Revolution.*
Floridi, L. 2010. *Information: A very short introduction*. Oxford: Oxford University Press.
Gelven, M. 1994. *War and existence*. Philadelphia: Pennsylvania State University Press.
Lauterpacht, H. Ed. 1952. *Oppenheim, international law.*
Libicki, M. 1996. *What is information warfare?* Washington, DC: National Defense University Press.
Matthias, A. 2004. The responsibility gap: Ascribing responsibility for the actions of learning automata. *Ethics and Information Technology* 6:175–183.
Saxena, K. B. C. 2005. Towards excellence in e-governance. *Journal of Public Sector* Management 18:498–513.
Schmitt, M. N. 2008. *Cyber operations in international law: the use of force, collective security, self-defense, and armed conflicts.* Proceedings of Workshop on Deterring Cyberattacks, National Research Council: The National Academies Press.
Sparrow, R. 2007. Killer Robots. *Journal of Applied Philosophy* 24:62–77.
Steinhoff, U. 2007. *On the ethics of war and terrorism*. New York: Oxford University Press.
Taddeo, M. 2012. Information warfare: A philosophical perspective. *Philosophy and Technology* 25:105–120.
Toffler, A., and H. Toffler 1997. Foreword: The new intangibles. In *In Athena's camp: Preparing for conflict in the information age*, eds. J. Arquilla and D. Ronfeldt, xii–xxiv. Santa Monica: RAND Corporation.

Contents

Part III Information warfare and virtue

Contributors

Keith Abney Department of Philosophy, California Polytechnic State University, San Luis Obispo, USA

Fritz Allhoff Department of Philosophy, Western Michigan University, Kalamazoo, USA

Deen Chatterjee University of Utah, Salt Lake City, USA

Dorothy E. Denning Department of Defense Analysis, Naval Postgraduate School, Monterey, USA

Randall R. Dipert Department of Philosophy, University at Buffalo, Buffalo, NY, USA

Don Howard Department of Philosophy and Reilly Center for Science, Technology, and Values,University of Notre Dame, Notre Dame, USA

Steven P. Lee Hobart and William Smith Colleges, Geneva, USA

Patrick Lin Department of Philosophy, California Polytechnic State University, San Luis Obispo, USA

George R. Lucas Ethics and Public Policy, Naval Postgraduate School, Mile Drive, 225-17, Pacific Grove, CA, USA

Wayne McCormack University of Utah, Salt Lake City, USA

Brian Orend University of Waterloo, Waterloo, Canada

Thomas W. Simpson Blavatnik School of Government, University of Oxford, Oxford, UK

Bradley J. Strawser Department of Defense Analysis, Naval Postgraduate School, Monterey, USA

John P. Sullins Sonoma State University, Rohnert Park, CA, USA

Mariarosaria Taddeo Cyber Security & Ethics, Department of Politics and International Studies, University of Warwick, Coventry, United Kingdom

Uehiro Centre for Practical Ethics, University of Oxford, Oxford, United Kingdom

Shannon Vallor Santa Clara University, Santa Clara, CA, USA

Part I
Analysing Information Warfare

Chapter 1
Fog in the Fifth Dimension: The Ethics of Cyber-War

Brian Orend

Abstract Cyber-warfare is a cutting-edge topic in armed conflict. It can be defined, at least initially, as attempting to use the Internet, and related advanced computer technologies, to substantially harm the fundamental interests of a political community. And cyber-space has been referred to as "the fifth dimension of warfare," after: land; water; air; and space. Yet, much confusion (or "fog") surrounds cyber-warfare, both regarding its present realities and its future potential. How much damage can cyber-attacks actually do? Is it even appropriate to liken computer-based cyber-attacks to physical ("kinetic") violence? Is "informational warfare", as cyber-war is otherwise known, changing the very nature of political conflict in our time (indeed, for all time)? This chapter aspires to clear up some—but certainly not all—of this fog which surrounds the fifth dimension. It will do so by means of critically examining three important distinctions in this regard. But first, some workable definitions are required.

> *All action takes place, so to speak, in a kind of twilight which, like a fog or moonlight, often tends to make things seem grotesque and larger than they really are.* (Clausewitz 1995)
> Carl von Clausewitz, ***On War***

Cyber-warfare is a cutting-edge topic in armed conflict. It can be defined, at least initially, as attempting to use the Internet, and related advanced computer technologies, to substantially harm the fundamental interests of a political community. And cyber-space has been referred to as "**the fifth dimension of warfare**," after: land; water; air; and space (The Economist 2010). Yet, much confusion (or "fog") surrounds cyber-warfare, both regarding its present realities and its future potential. How much damage can cyber-attacks actually do? Is it even appropriate to liken computer-based cyber-attacks to physical ("kinetic") violence? Is "**informational warfare**", as cyber-war is otherwise known, changing the very nature of political

B. Orend (✉)
University of Waterloo, Waterloo, Canada
e-mail: bdorend@uwaterloo.ca

L. Floridi, M. Taddeo (eds.), *The Ethics of Information Warfare*,
Law, Governance and Technology Series 14, DOI 10.1007/978-3-319-04135-3_1,

conflict in our time (indeed, for all time)? This chapter aspires to clear up some—but certainly not all—of this fog which surrounds the fifth dimension. It will do so by means of critically examining three important distinctions in this regard. But first, some workable definitions are required.

1.1 Quick Definitions

As offered above, "**cyber-warfare**" is an umbrella term, referring to the aggressive use of advanced computer technologies in a way deliberately designed to substantially harm the fundamental interests of a political community (Carr 2010). A **political community** can be considered a country, state, or nation, but was perhaps most suggestively defined by Aristotle as an on-going human partnership, formed for the sake of the common good of its members and aimed at achieving both *justice for all* and *happiness for each* (Aristotle 1984). Political communities have many interests, but among the most fundamental are: peace and security; access to vital resources; the right to govern themselves free from foreign domination; the right to grow their economy and try to improve their lives; and finding a balance between creative innovation and reliable stability in their way of life. Most basic, perhaps, of a country's **fundamental interests** are: freedom from invasion; freedom from domination; and secure possession of those resources truly needed to survive on an on-going basis (Orend 2013).

Cyber-warfare, generally, can take one of three forms:

1. **espionage** (i.e., using the Internet, etc., to gather information which a country has taken steps to protect as a matter of national security, such as secret-, confidential-, or classified information);
2. **the spread of disinformation**, via the same means, in a manner which harms the security interests of the target country; and/or
3. **sabotage** (i.e., using these means to bring about the non-functioning, or destruction, of various systems which are integral to the basic interests of a political community. The systems most often mentioned include: electricity and power; water and fuel distribution; computerized parts of manufacturing facilities; transportation systems, such as air or rail; banking and the stock market; and even the Internet itself, or at least the most used web-sites (like Google or Facebook), Internet service providers, and/or the most basic operating systems.) (Clarke 2010)

Cyber-attacks would then refer to any *specific* use of any of 1–3 above, as tools within the overall cyber-warfare *strategy*. The countries most frequently mentioned today with reference to cyber-war technology include: America; Britain; China; France; India; Israel; Pakistan; and Russia (Clarke 2012).

1.2 The First Distinction: "Cyber-War-Skeptic" Vs. "Cyber-War-Salesman"

A **cyber-war-skeptic** would be someone, like Howard Schmidt, who declares that "there is no such thing as cyber-warfare." (Schmidt 2006) A cyber-war-skeptic believes that the threat from such measures (as 1–3 above) is minimal or, at least, not at all on a level where talking about a military response is appropriate. A cyber-war-skeptic might also believe that the whole extended analogy—between information attacks and computer viruses, on the one hand, and kinetic warfare and physical casualties, on the other—is: 1) crude and factually incorrect, perhaps even a "category mistake" confusing two completely different things; 2) fear-mongering, capitalizing on the common person's (relative) intimidation by, and lack of knowledge regarding, advanced computer technology; and 3) deliberate exaggeration, or even fraud, communicated by those with a vested interest in the business of cyber-security, ranging from cash-strapped military departments looking for fresh resources to greedy software programmers drooling at the prospects for profit. (And the financial stakes *are* very considerable: The Pentagon has publicly disclosed that, in the first half of 2009 alone, it spent over $ 100 million USD "responding to, and repairing damage from, cyber-attacks.") (Clarke 2010).

A **cyber-war-salesman**, on the other hand, would be someone who wildly exaggerates the threat of cyber-war, and the disruption to be suffered from such. It needs to be stressed that such a figure doesn't have to be a cyber-war profiteer, as just mentioned at the end of last paragraph. Consider that the influential 2010 *Lipman Report*—i.e., the US Congress' formal study of cyber-warfare, for American foreign policy purposes—warned that threats of "crippling attacks on computer networks are sharply on the rise." (U.S. Congressional House 2011) The US mainstream broadcasting company CBS reported, in an evening national TV broadcast, that, in 2007, the US federal government suffered "an espionage Pearl Harbour" when some unknown sources downloaded "terabytes of classified government and even military information." (Note the similarity between the spoken sound of "terabytes" and "terror bites.") (CBS News 2009) Indeed, in 2010, the U.S. Joint Forces Command issued a statement expressing its conviction that "adversaries have already taken advantage of computer networks and the power of information technology… to plan and execute savage acts of terrorism." (USJFC 2010) Even the normally staid *New York Times* reported that a **malware program** (i.e., a malicious software virus) which had infected some U.S. factory computers should be "considered the first attack on critical industrial infrastructure that sits at the foundation of modern economies." (New York Times 2010) Finally, consider the closing lines in one of distinguished journalist Michael Gross' important articles about information warfare in general, and a virus called "Stuxnet" (more below) in particular:

> Cyber-conflict makes military action more like a never-ending game of uncle, where the fingers of weaker nations are perpetually bent back. The wars would be secret, waged by members of anonymous, elite brain trusts, none of whom would ever have to look an enemy in the eye. For people whose lives are connected to the targets, the results could be as cata-

> strophic as a bombing raid, but would be even more disorienting. People would suffer, but would never be certain whom to blame.
> Stuxnet is the Hiroshima of cyber-war. That is its true significance, and all the speculation about its target and its source should not blind us to that larger reality. We have crossed a threshold, and there is no turning back. (Gross 2011a)

Within these various dramatic comments, one takes special note of the multiple references to terrorism and to the Second World War.

So, the question naturally arises: who's right? Who, between the cyber-war skeptic and the cyber-war salesman, is more correct? It will be an on-going theme of this chapter that the truth between each of the three distinctions to be drawn and addressed probably rests somewhere in the middle, in a Rawlsian over-lapping consensus (as it were) (Rawls 1993). Where does the middle ground properly lay in this present case?

1.2.1 Middle Ground Judgment

On the one hand, there is no denying that cyber-attacks *are* real, and they have had some surprisingly serious consequences, at times very much akin to actual, kinetic warfare. So, in this sense, the cyber-war skeptic is wrong and the cyber-war salesman, right. Several quick, illustrative examples:

- In 1982, during the height of The Cold War, a Canadian oil and gas company thought they had a Soviet (Russian) spy in their midst. They contacted America's military. The Canadians and Americans launched a joint scheme: they would let the spy steal what he was after: a computer-control system for regulating the flow of oil and gas. (The Russians wanted this to modernize their pipeline system in Siberia.) But the Americans programmed the computer system with "a logic bomb", designed to make the pipelines malfunction and eventually explode after it was implemented. And that is exactly what happened, with some loss of life and a substantial set-back for a key sector of the Soviet economy (The Economist 2010).
- In 2007, Russia launched a cyber-attack on Estonia, a neighbouring country. There was a dispute between them regarding the movement of a war statue of great meaning to the Russians. When the Estonians moved it, Russia responded with a crippling cyber-attack on the websites of the Estonian government, media, and its richest banks. For nearly a week, these institutions could not conduct any business online, nor could their citizens/customers contact them, or access anything through them. The attack came to an end only when Russia decided to release its grip (Karatzogianni 2008).
- From May to December, 2010, India and Pakistan traded over 1,000 separate cyber-attacks against each other, directed not only against official government- and military web-sites but also selected high-profile companies, universities, and research institutes. While most of these attacks were mere "**defacements**" of the various web-sites (and thus more a form of disinformation, or graffiti, than

sabotage), they nevertheless revealed both the involvement of these countries in cyber-war activities as well as the degree to which they were capable of gaining access and demonstrating control (Hyacinthe 2011).

- More seriously, in 2010, Iran was attacked by a computer virus or "worm" commonly believed to have been the joint-creation of both America and Israel (nicknamed "**Stuxnet**"). A piece of malware, this very sophisticated computer virus was planted in a German-made component of one of Iran's nuclear reactors. When it was activated, the virus eventually disabled the reactor, forcing it to shut down—lest it melt-down and cause enormous damage—for an unspecified time (thought to be at least for months, and perhaps even over 1 year). The goal, reputedly, was to set-back Iran's progress towards developing nuclear weapons (Gross 2011a).
- Perhaps relatedly, in 2012 the "**Flame**" virus entered public knowledge. Reputedly, Flame went undetected for over 5 years. Its main purpose seems to have been espionage or information-gathering. Experts have pronounced it "more than 20 times more powerful" in its sophistication than Stuxnet and, by time of discovery, it was confirmed to be present in over 5,000 computers, almost all in the Middle East, with a special concentration in Egypt, Iran and Israel (Stallwood 2012).
- The country most associated with cyber-attacks today is China. Unlike American and Russian attacks, though, which have tended to feature sabotage, the Chinese seem to prefer espionage, both of the commercial- and political variety. Many of the top US high-tech firms, such as Google, Microsoft, Apple, and various weapons companies, have complained of sustained cyber-attacks from China which have accessed tons of their highest-security information, including especially product design- and patent information (as well as, intriguingly, human resources data, such as personal information about top executives). The companies have pressed the US government to respond, but thus far all that have been issued are verbal warnings by Hilary Clinton, the US Secretary of State (Gross 2011b).

Thus, informational warfare is truly real; and it can have—and has had—very serious consequences, including loss of life. (Though, admittedly, these most serious consequences seem more rare and exceptional rather than regular and expected, as with direct kinetic warfare.) On the other hand, the cyber-war skeptic seems correct to insist that it's important not to exaggerate people's fears about the likelihood of themselves being victimized by such strikes, or to make colourful but unhelpful analogies to weapons (such as at Hiroshima) which can kill hundreds of thousands of people. And it certainly seems compelling to note that all this talk, and all this activity, surrounding cyber-warfare *does serve some vested interests*, out to gain narrow advantage, and we should regard their claims with some sober second thought, and make them prove such. After all, if The Pentagon is spending $ 100 million USD every 6 months on cyber-defence, one must admit that a pot of money *that* large is likely to attract not only legitimate, but also questionable, attention.

1.3 The Second Distinction: Realism Vs. Just-War Theory

1.3.1 No Law

Cyber-warfare is here, and it's real; and so the question arises: what, if anything, should we do about it? An obvious response would be to try to regulate it with the law. Presently, there is no international law whatsoever regarding informational warfare. In 2011, America, China and Russia got together for a high-level meeting of officials, one branded as being "talks about talks" regarding a possible negotiated treaty between them on the acceptable methods and means of cyber-war (analogous to the many such treaties on kinetic warfare). But the talks fell apart, amidst bitter mutual accusations (Dinniss 2012).

It is vital to note that, in the absence of an international treaty on this, all the major countries have simply (and resoundingly) declared that, as a matter of their foreign policy, they will consider any "severe" cyber-attack against them as a ***casus belli***, i.e.: a cause for war, a reason to resort to war (presumably, either of the new informational-, or the traditional physical, kind) (Carr 2010; Lynn 2010).

1.3.2 Thus, Ethics

In the absence of law, one turns to ethics for guidance. Traditionally, there are three major traditions of thought about the ethics of war and peace: realism; just war theory; and pacifism.

1.3.3 Pacifism

1.3.3.1 In General

Pacifism is a species of idealism regarding international affairs. **Idealism** is the view that one's goal as a country, when dealing with others, ought to be to *do one's part in making the world a better place*. It's like a form of national altruism, or unselfishness. When dealing with the outside world, use one's resources and influence to improve the world: make it richer, happier, more secure, and so on. Be a good international citizen. Give a damn, so to speak, and act accordingly. Commonly, idealists tend to divide into those favouring small-scale, concrete, and gradual improvements versus those attracted to larger-scale, sweeping, and more sudden shifts in international politics. Prominent idealist thinkers would include Immanuel Kant, whereas prominent idealist politicians would include former US President Woodrow Wilson (Orend 2013; Price 2007; Kant 1983).

1.3.3.2 On War and Cyber-War

The essence of **pacifism**, obviously, is a rejection of war. War is always wrong; there is always some superior alternative to war, such as non-violent resistance. For better or worse, pacifism thus far plays no effective role in the debate about cyber-warfare. While there *has* been a pacifist-inspired idea to have a treaty banning all forms of cyber-war, and to have the Internet declared "part of the common cultural heritage of humanity" (and thus, *not* a proper battlefield), it has thus far gone nowhere (Menthe 1998). (Mathieu Doucet has suggested that perhaps pacifists could actually endorse cyber-warfare, as a tool of nonviolent resistance designed to shut down, or seriously complicate, the use of kinetic war, with its violent, bloody consequences. That is an intriguing idea but one which, thus far, remains underdeveloped—and, further, it would need to account for, and ethically grapple with, those forms of cyber-war where killing force *has* resulted, such as in the Soviet logic bomb case described above. Pacifists, presumably, could neither logically nor morally endorse those forms: care would need to be taken to show which kinds of cyber-war might be permitted and which not.)

1.3.4 Realism

1.3.4.1 In General

Realism is the view that, as a country, one's goal should be *to advance one's national interests*. **National interests** are those things that improve, benefit, or enhance the position of one's country. They boil down to having both hard- and soft power. **Hard power** is the use of economic resources, and/or armed force, to get what one wants in international relations. Summarized as "the bucks-and-bullets" approach to foreign policy, it means either buying or forcing the compliance of others to one's will. **Soft power**, by contrast, is the use of one's language, ideas, values, and culture to bring about the compliance of others to one's will. The spread of one's culture is thought to create a commonality of world-view, a mutuality of interest, and a reservoir of good will, which bolsters one's ability to get what one wants. Realism is thus like a form of national egoism or selfishness. When dealing with the outside world, or "the international community," one ought to (as they say) "Look Out For Number One." Do the best one can for one's own society, especially in terms of: national security and defence; growing the economy, optimizing one's population and its access to natural resources; and augmenting one's cultural and political influence around the world. Prominent realist thinkers would include Machiavelli and Hans Morgenthau. Prominent realist politicians would include Henry Kissinger and former US President Richard Nixon (Orend 2013; Machiavelli 1998; Kissinger 1995; Morgenthau 1970).

1.3.4.2 On War and Cyber-War

In terms of warfare, realism clearly clashes with pacifism. Indeed, whereas pacifism seems to say "nothing goes" in terms of violent warfare, realism replies with an equally sweeping "anything goes." The most important thing to note with warfare, according to many realists, is that history shows that it is a dreadful thing to lose a war, and such is almost never in the interests of any country. Thus, the over-riding objective is to *do whatever one deems required to win*. It all becomes a calculus of national self-interest and advantage.

The realist thus views the development of informational warfare as just the latest permutation in human history's endless cycle of violent conflict. Countries constantly seek advantage in war, including especially through mastery of new weapons. Cyber-technology is a new, and potentially very important, technology which could have deep implications for armed conflict, in particular, and for the hierarchy of nations in general. Thus, countries *should* be doing exactly what they *are* doing right now: investing in its development; experimenting with its use; not tolerating any strikes against themselves; threatening others over its use; using it against others to gauge its impact; and ascertaining how best to use this new weapon in their overall war-fighting, and foreign policy, objectives.

1.3.5 Just War Theory

In-between the extreme views of realism and pacifism resides just war theory. Like pacifism (and unlike realism), just war theory believes that there *is* both sense and value in applying ethics and moral values to issues of international relations. But unlike pacifism (and like realism), just war theory believes that there *can sometimes* be instances where resorting to war is justified, if only as "the least-worst" option. Thus, if pacifism says "nothing goes" with regard to the ethics of war, and realism declares that "anything goes", just war theory opines that "something, sometimes goes." While war *can* be morally permissible, just war theory nevertheless views war dimly and dangerously, and insists that it's too risky and lacking in restraint to allow for "anything goes." Just war theory seeks to substitute, for that realist permissiveness, a set of sensible rules to restrain and guide those considering warfare as a tool for solving some serious foreign policy problem. The just war approach has been deeply influential on the international laws of armed conflict, for instance as contained in the *Hague*- and *Geneva Conventions*, as well as in the *UN Charter* and the various resolutions of the UN Security Council (UNSC) (Walzer 1977; Orend 2006; Roberts and Guelff 1999).

Elsewhere, I've further explained, and defended at length, the claims and rationale of just war theory against its two major rivals. I still believe it is the most sophisticated, detailed, comprehensive, and well-defended system of thought about the ethics of war and peace (Orend 2006, 2000a, 2009). As such, it profits us to consider further *how* the just war rules and categories can shed light on informational warfare. The method will be as follows: to explain just war rules in general (as they've applied traditionally to regular, physical warfare); and then to suggest how they might apply to our analysis of informational warfare.

1.3.5.1 Jus ad Bellum

This is Latin for "the justice of war." When, if ever, may states fight?

The just war answer is that states may fight *only if* they satisfy *all* of the following rules: just cause; right intention; public declaration of war by a proper authority; last resort; probability of success; and proportionality. Those with "the war power" (usually the executive branch in non-democratic societies, and the legislative branch in democratic ones) are to ensure they satisfy these principles before embarking on war.

- **Just Cause**

The way international law renders just war theory in this regard is very clear and quite helpful. Most experts agree that, when it comes to a just cause for war, three general principles are at play:

1. All countries have the inherent, or "natural," right to go to war in **self-defence** from aggression. **Aggression** is defined as any unjustified use of force against another country. Any armed attack which crosses an international border constitutes aggression and is a *casus belli*, i.e., "a cause for war."
2. All countries have the further natural, or inherent, right of **other-defence**—otherwise known as "**collective security**"—to go to war as an act of aid, or assistance, to *any* country victimized by aggression; and
3. *Any other* use of force—e.g., pre-emptive strike, or armed humanitarian intervention—is *not* an inherent, or natural, right of states. Any country wishing to engage in such is supposed to get *the prior approval* of the UNSC. Failing to receive such prior authorization renders any such use of force illegal, itself an act of aggression (Orend 2006, 2013; Regan 1996; Roberts and Guelff 1999).

So, if Country A commits an armed attack against Country B, then B (and any other country C) is entitled to go to war against A as an act of *defence from*, *resistance to*, and *punishment of*, aggression. Aggression is seen as a wrong so severe that war is a fitting response because it violates the most basic rights of groups, and individuals, to life and security, and to freedom and well-being—i.e., to go about their lives peacefully, on a territory where their people reside. Classic examples of international aggression include: Imperial Germany's invasion of Belgium in 1914, sparking WWI; Nazi Germany's invasion of Poland in 1939, sparking WWII; Japan's invasion of China in 1937, and its attack on the USA at Pearl Harbour in 1941, sparking the Pacific part of WWII; the USSR's invasion of Afghanistan in 1979; and Iraq's invasion of Kuwait in 1990, sparking the Persian Gulf War. There are actually thousands of historical examples of international aggression (Orend 2006, 2013; Walzer 1977; Keegan 1990, 1994).

- **Proportionality**

In every kind of law or rule, there is supposed to be a **proportion**, or balance, between problem and solution (or between violation and response), which is here to say that international law commands that the problem in question really must be so serious that war is a proper reply. Since war is so costly, bloody, and unpredictable, it follows that only a very few problems in international life are truly so bad that war will be a proportionate response to them. The function of this rule is to get those with the war power to think again, deeply, whether there isn't some other thing to be tried—say, one of the other foreign policy tools, such as diplomacy or sanctions—before resorting to force. What, if anything, might be a problem truly so severe that war is a proportionate response? *The answer of international law, and just war theory (for reasons stated above) is: aggression.* When confronted with an aggressive invader—like Nazi Germany, Imperial Japan, or the Soviet Union—who's intent on conquering and essentially enslaving other nations, it's deemed reasonable to stand up to such a dark threat to life and liberty and to resist it, and beat it back, with force if need be. Just as dangerous criminals must be resisted and not be allowed to get away with their crimes, countries are entitled to stand up to aggressors, and to resist and defeat them (Orend 2006; Walzer 1977).

- **Public Declaration of War by a Proper Authority**

War is supposed to be declared out in the open, officially and honestly, by the proper authority for doing so. In every country, some branch of government has "**the war power**:" i.e., the authority to order the use of force and warfare. In Canada and Britain, the war power rests with Parliament; in America, the war power likewise rests with the legislature: i.e., Congress. But the American President—as Commander-in-Chief of the Armed Forces—has enormous factual power to order the American military into action. As a result, many experts argue that the war power in the US is actually split—in classic American "**checks-and-balances**" style—between the legislative and executive branches of government. (This became an issue of struggle between the branches during both the Korean War (1950–1953) and especially the Vietnam War (1954–1974), when Congress felt successive presidents were running a ***de facto*** war without actually publicly declaring it and getting ***de jure*** authority for doing so—i.e., getting a clear vote of support from Congress.) (Regan 1996) Generally, in most democracies, the legislature has the war power whereas, in most non-democratic societies, it's the executive—i.e., the president or dictator—which has the authority to order war. We have seen, further, how in all cases where non-defensive armed force is being considered, the UNSC must also approve of the action, and beforehand. This is to say that, with non-defensive war, both domestic and international authorization must be satisfied (Orend 2006, 2013).

- **Last Resort**

State governments are only supposed to go to war as a **last resort**, only after all other reasonable means of problem-solving have been tried, and failed. It's said that countries have four basic tools in their foreign policy tool-box: diplomacy; economic incentives; sanctions; and force. Obviously, you want to exhaust all other means of problem-solving before engaging in something as expensive, bloody, and risky as war. A nice illustration of this rule in action happened during the run-up to the Persian Gulf War of 1991. In August 1990, Saddam Hussein's Iraq invaded its tiny neighbour, Kuwait. International allies, as led by the USA and UK, tried to talk to Saddam and threaten him, to no avail. They then slapped sweeping sanctions on him, and got most of his neighbours to agree and also put pressure on Iraq. Still nothing. As a result, the international community felt it was the last resort to go to war to push Saddam out of Kuwait, and back into his own borders. This they did, within 2 months, in early 1991 (Johnson and Weigel 1991; Orend 2006).

The above *jus ad bellum* rules are all part of the international laws of armed conflict. Just war theory, as a theory of ethics, levies two additional moral requirements:

- **Right Intention**

The notion here is that one's motives need to be ethically proper. It's not enough merely that one's *actions* comply with the above rules but that, furthermore, one acts with *the right frame-of-mind* and, in particular, that seedy, ulterior motives—such as greed—play no role. In the case of a just war, then, the idea would be that one's intentions in acting are to resist, repulse, and punish aggression, and nothing more. Though this rule is *not* part of international law—largely owing to the difficulty involved discerning the true intentions of a complex, multi-part actor like a state government—it is frequently invoked in common moral discussion of warfare. It was, e.g., a popular criticism of the Bush Administration's decision to invade Iraq in 2003 to suggest that the decision had as much, or more, to do with the desire to gain secure access to oil as it did with, say, ensuring Iraq wasn't about to deploy weapons of mass destruction (WMD) against the USA (Murray and Scales 2003; Woodward 2004; Orend 2006).

- **Probability of Success**

The rule here is that one should not begin a war one knows in advance is going to be futile. The point is *to prohibit pointless killing and suffering*: one should have some probability of success before resorting to war. At the same time, this can be very difficult to predict at the start of war, and history has shown that, sometimes, long-shots can actually win. Moreover, this rule seems biased in favour of powerful states, who (for that very reason) have better chance of winning their wars. This probably explains the absence of this rule from international law, which is based around theoretical ideals regarding the equality of sovereign states: if a country—any country, big or small—has been victimized by aggression, who are we to say that they shouldn't go to war, because at the outset it looks like such a risky venture? (Orend 2006)

1.3.5.2 Application of Jus ad Bellum Rules to Cyber-War

- **Just Cause**

As shown above, the gold standard of *casus belli* is a kinetic physical attack, usually involving some kind of armed invasion across a border. As such, a cyber-strike does *not* seem to constitute aggression in the traditional meaning of the term. But two thoughts suggest themselves:

1. sometimes, as we've seen, cyber-attacks can actually lead to traditional, physical damage, including loss of life. Such would have to be construed as straight-forward instances of aggression, ethically and legally enabling a forceful response.
2. an argument could be made that the concept of aggression itself needs to be amplified and expanded (but responsibly so) precisely to allow for cyber-attacks as a kind of aggression. This thinking would need to stress the new and pervasive role which advanced computer technologies have come to play in our lives (especially in the developed world), and the degree to which damage aimed at them could rise to the level of a very serious, society-wide strike. Indeed, some might even argue that, say, a cyber-attack on the stock market, causing it to crash and costing millions of people billions of dollars, might actually be more damaging in long-term consequence than, e.g., an army unit lobbing a missile across a border, resulting in the physical injury (or death) of only, say, 3 soldiers on border patrol. This is to say that: a) a powerful cyber-strike might actually be more damaging than a physical strike; and b) if the latter counts as aggression, then the former ought to, as well. The key notion here would be that our thinking of what constitutes aggression needs to keep pace with the times and the new technological realities of our lives (Floridi 2010a, b).

I think these reflections would need to be made in greater detail than here, but I do support the general notion that these concepts, to remain relevant, must be considered in light of the latest technologies and deep, ongoing developments in the contours of our lives. My own considered view is that *a cyber-strike probably will not justify anything more than an in-kind cyber-response*, and that the burden of proof rests on anyone arguing that it may justify something further, such as an armed kinetic attack in reply. While there is much defensive, deterrent-based wisdom in the status quo—i.e., of warning others that any "severe" cyber-strike will be considered a *casus belli* (especially one involving sabotage against core, society-wide, infrastructure)—more sustained efforts at deeply developing these concepts need be made (Brenner 2009; Cook 2010; Lucas 2011).

- **Proportionality (and Probability of Success)**

Proportionality would clearly support the notion that a cyber-strike probably justifies *only* an in-kind cyber-response, and *not* an armed kinetic war in reply. And probability of success demands that we ask, for any such cyber-strike: is it likely to achieve its aims? How so? What kind of confidence can one have in that regard, especially as regards the minimization of any over-spill onto civilians and the likelihood that one can have favourable control over the consequences?

- **Last Resort**

There is a real danger, and some evidence from the actual uses thus far, that a major temptation with cyber-strikes is that they be used *not* as a last resort, but rather as a **first-strike capability**, either on their own or else to disorient and "soften up" the target for an actual kinetic attack, such as with drones, missiles, or even an armed invasion.

It might be argued that cyber-war could be rendered consistent with this principle, and find its own proper slot in the moral hierarchy of foreign policy tools, with diplomacy at the ground floor, as the most accessible (and encouraged) level, and with kinetic force at the top: the rarest, riskiest, and most controversial. Cyber-strikes could be located either just beneath kinetic warfare, or else perhaps on par with sweeping economic sanctions, which often are similarly targeted at foundational aspects of the target country's economy.

- **Public Declaration by a Proper Authority**

Here there is no question that the vast majority of actual cyber-attacks thus far have violated this rule of just war. Indeed, has any government publicly declared, and accepted responsibility, for any cyber-attack? One of the seductions of this technology is its supposed anonymity (though, almost always, the doer's identity *does* come to be known: see more below). We know that, historically, those with the war power prefer to use it in secret and with few, or no, checks-and-balances on them. Cyber-war may thus provide terrible temptations in favour of "easy war" and "secret war" which ought, obviously, to be resisted.

Experts in the field talk repeatedly of "**the attribution problem**", noting how cyber-attackers—especially those suspected to be linked, in some way, with China—go out of their way to hide their tracks and conceal the ultimate source of the strike. This is of great concern, as it would no doubt colour our judgment of whom it is permissible to strike back at (Clarke 2010). Yet, while being ignorant of the sophisticated details of how these things get determined, I would want to point out, as mentioned above, that eventually—and rather quickly, actually—the cyber-community seems to have been able, thus far, to come with pretty reliable attributions. Is cyber-strike attribution really so different from, and so much more difficult than, say, the investigations which went into determining who was responsible for the 9/11 attacks (i.e., al-Qaeda), and how the then-government of Afghanistan was complicit in them as well?

1.3.6 Traditional Rules of Jus in Bello

Whereas the rules of *jus ad bellum* are aimed at those with the war power—often the head of state—the rules of *jus in bello* are aimed at soldiers and officers, i.e., those who actually do the fighting. If they violate these rules, they can find themselves—after the conflict—facing war crimes charges, either domestically through their own military justice system or internationally through The Hague. There are many rules of *jus in bello*, but most of them concern only physical, kinetic warfare, and they are not directly applicable to cyber-war. The one principle which most clearly is, though, is *jus in bello*'s most important: discrimination and non-combatant immunity. Let us consider this first in the traditional sense, and then the potential implication for information warfare.

- **Discrimination and Non-Combatant Immunity**

“**Discrimination**” here means the need for fighters to distinguish, or discriminate, between legitimate and illegitimate targets, and to take aim only at the former. A **legitimate target** is anyone, or anything, which is part of the war machine of the enemy society. “**The war machine**” refers to the military-industrial-political complex which guides the war and fights it. Loosely speaking, it is anything which is a source of potential physical harm, or armed force, directed against oneself. More specifically, legitimate targets include: soldiers, sailors, marines, pilots, and their officers; their weapons and equipment; their barracks and training areas; their means of transportation; their supply and communications lines; and the industrial sites which produce their supply. Core political- and bureaucratic institutions are also legitimate objects of attack, in particular things like the Defence Ministry. **Illegitimate targets** include residential areas, schools, hospitals, farms, churches, cultural institutions, and non-military industrial sites. *In general, anyone or anything not demonstrably engaged in military supply, or military activity, is immune from direct, intentional attack.* Thus, **non-combatants**—i.e., civilians—are “immune” from intentional attack. This is seen as probably the worst war crime: the intentional killing of civilians (Walzer 1977; Orend 2006).

Strange as it may sound, the non-combatant immunity principle does *not* mean that it’s illegal for civilians to die in wartime. What is illegal is *taking deliberate and intentional aim* at civilians with armed force. If a fighting side has taken every reasonable effort to avoid and minimize civilian casualties—but some civilians still die accidentally, or in the indirect way just noted—then that is *not* a war crime. Such civilians are viewed as “**collateral damage**”—i.e., accidental, un-intended, casualties of the fighting. An example would be an air-bombing raid on an enemy’s industrial sites, during which a few bombs accidentally go astray and hit a close-by residential area, wounding and killing some civilians.

So, civilians are *only* entitled to “due care” from fighters; they are *not* entitled to absolute and fail-safe immunity from warfare. What does “**due care**” include? It includes all serious and sustained efforts, from the top of the military chain of command down to the bottom, to protect civilian lives as best as can be amidst the difficult circumstances of war. So, e.g., strategists must make their plans with an eye to minimizing civilian casualties; intelligence needs to be gathered and analyzed regarding which are the permissible targets; soldiers need to be trained exhaustively in proper—i.e., restrained and discriminating—ways of fighting; and any rough treatment of civilians needs to be investigated and punished; and so on (Orend 2006; Walzer 1977).

What about so-called “**dual-use**” targets? The question arises: what about things used *both* by the military and civilians during war: e.g., roads, bridges, radio and TV networks and transmitters, railway lines, harbours, and airports? International law forbids targeting them but, in reality, they often are, as they are so useful in helping military planners communicate with their troops and to move them around to where they can fight. More controversial, and thus more criticized, is targeting basic infrastructure, like farms, food supply, sewers, water treatment plants, irrigation systems, water pipelines, oil and gas pipelines, electricity generators, and power and telephone lines. The civilian population pays a huge price for any damage inflicted on such vital social infrastructure, and so it seems to violate civilian immunity to go after them. America did this recently twice. During the opening days of both the 1999 Kosovo War, and the 2003 Iraq attack, America launched a so-called "**shock and awe**" campaign—relying on air power, bombing raids, and cruise missiles—to inflict heavy damage on basic infrastructure (especially communications and electricity) on Serbia and Baghdad, respectively. The military goal of such a strike is to hit the enemy as fast and furiously as possible, dazing them, and "softening them up" for a subsequent ground invasion by army soldiers. It is also to shock the civilians in that society into putting pressure on their regime to give up and surrender quickly (Clark 2002; Ignatieff 2001; Orend 1999, 2006).

1.3.6.1 Application of Jus in Bello to Cyber-War

The inference for cyber-warfare is clear: if one engages in a cyber-strike, one ought to take every effort to ensure that civilians are left out of it, and that only legitimate targets bear the brunt of the cyber-attack. The best, contrasting examples from the above list of cases would be the Russian cyber-strike on Estonia, on the one hand, and Stuxnet, on the other. The Russian strike clearly impacted every citizen in Estonia, as for the week or so in which it was on-going, such citizens could not have contact with their democratically-elected government online, nor could they access personal funds from their own bank accounts, and so on. This, clearly, was a substantial and intended interference with the basic rhythms of their daily lives. Ironically, Estonia (and the other Baltic states) had been, up until that point, at the fore-front of so-called "**e-government**": i.e., making as many government services deliverable over the Internet as possible. The cyber-strike from Russia, unfortunately, showed the potential disadvantages of such a progressive and technologically advanced approach. In any event, it clearly violated non-combatant immunity.

Stuxnet, by contrast, was elaborately constructed to harm only the nuclear power capability of the Iranian government. And it seems to have succeeded in that regard, and not one civilian was even harmed—much less killed—in the process. (The virus, after it struck, was programmed to "evaporate;" i.e., write itself out of existence so it could do no further harm.) Now, I suppose one could talk about the *potential* harm to the public, had the Iranians not known how to handle the situation: things may, indeed, have taken a frightening turn. Obviously, the perpetrators (rumoured to be the US and Israel) had confidence that the Iranians *would* recognize what was happening, and would have the wherewithal to shut the reactor down and not risk broader public damage. In any event, these two broad examples show what just war theory would view as a permissible cyber-strike: a discriminate one aimed only at a legitimate target, and with clear measures taken to minimize or eliminate any negative consequences on civilian populations. Especially to be ruled out—as the equivalent, really, of WMD—are potent, society-wide, cyber-strikes involving sabotage of basic core infrastructure (like, say, water treatment) seeing as how such would predictably involve large-scale damage, harm, and loss of life (Lucas 2011, Cook 2010).

1.3.7 Jus Post Bellum: The Aftermath of War

The final phase of war is when the conflict is coming to an end. *Jus post bellum* concerns "justice after war." There is, perhaps surprisingly, very little international law regulating things in this regard. The preference, historically, has been for "the winner to enjoy the spoils of war:" i.e., for the war winner to impose whichever terms of peace it prefers upon the loser (Orend 2000a, 2002b). Generally, one of two approaches tends to be followed in this regard: retribution or rehabilitation.

- **Retribution**

According to **the retribution model,** the basic aspects of a decent post-war peace are these (and, crucially, they assume that "the good side" won, and that the aggressive side lost):

- A public peace treaty.
- Exchange of Prisoners of War (POWs).
- Apology from the Aggressor.
- War crimes trials for those responsible.
- Aggressor must give up any gains made during the war.
- Aggressor must be demilitarized, at least to avoid a repeat.
- Aggressor must suffer further losses. What makes this model one of **retribution** is the conviction that it is *not enough* for the defeated aggressor merely to give up what it wrongly took, plus some weapons. *The aggressor must be made worse off than it was prior to the war.* Why? The defenders of this model suggest several reasons. First, it is thought that justice itself demands retribution of this nature—the aggressor must be made to feel the wrongness, and sting, of the war which it unjustly began. Second, consider an analogy to an individual criminal: in domestic society, when a thief has stolen a diamond ring, we don't just make him give the ring back and take away his thieving tools. We also make him pay a fine, or send him to jail, to impress upon him the wrongness of his conduct. And this ties into the third reason: by punishing the aggressor, we hope *to deter or prevent* future aggression, both by him (so to speak) and by any others who might be having similar ideas.

But what will make the aggressor worse off? Demilitarization, sure. But two further things get frequently employed: *reparations payments* to the victims of the aggressor, plus *sanctions* slapped onto the aggressor as a whole. These are the post-war equivalent of fines, so to speak, on all of the aggressive society. Reparations payments are due, in the first instance, to the countries victimized and hurt by the aggressor's aggression and then, secondly, to the broader international community. The reparations payments are *backward-looking* in that sense, whereas the sanctions are more *forward-looking* in the sense that they are designed to hurt and curb the aggressor's future economic growth opportunities, at least for a period of time (a sort of probation) and especially in connection with any goods and services which might enable the aggressor to commit aggression again (Orend 2002b, 2006).

- **The Rehabilitation Model**

There is no sharp split between the retribution and rehabilitation models. They share commitment to the following aspects of a decent post-war settlement: the need for a public peace treaty; official apologies; exchange of POWs; trials for criminals; some demilitarization; and the aggressor must give up any unjust gains. Where the models differ is over three major issues. First, the rehabilitation model *rejects sanctions*, especially on grounds that they have been shown, historically, to harm civilians and thus to violate discrimination. Second, the rehabilitation model *rejects compensation payments*, for the same reason. In fact, the model favours *investing in* a defeated aggressor, to help it re-build and to help smooth over the wounds of war. Finally, the rehabilitation model *favours forcing regime change* whereas the retribution model views that as too risky and costly. That it may be, but those who favour the rehabilitative model suggest that it can be worth it over the long-term, leading to the creation of a new, better, non-aggressive, and even progressive, member of the international community. To those who scoff that such deep-rooted transformation simply can't be done, supporters of the rehabilitative model reply that, not only *can* it be done, it *has* been done. The two leading examples are West Germany and Japan after WWII (Orend 2000a, 2006).

Based on these best-case practices (Dobbins et al. 2003; Dobbins and Jones 2007), supporters of rehabilitation have devised their own list of desirable elements during the post-war period. The occupying war winner, during post-war reconstruction, ought to:

- *Adhere diligently to the laws of war during the regime take-down and occupation.*
- *Purge much of the old regime, and prosecute its war criminals.*
- *Disarm and demilitarize the society. (But then:)*
- *Provide effective military and police security for the whole country.*
- *Work with a cross-section of locals on a new, rights-respecting constitution which features checks and balances.*
- *Allow other, non-state associations, or "civil society", to flourish.*
- *Forego compensation and sanctions in favour of investing in and re-building the economy.*
- *If necessary, re-vamp educational curricula to purge past propaganda and cement new values.*
- *Ensure that the benefits of the new order will be: (1) concrete; and (2) widely, not narrowly, distributed.*
- *Follow an orderly, not-too-hasty exit strategy when the new regime can stand on its own two feet* (Orend 2006).

1.3.7.1 Application of Jus post bellum to Cyber-War

It's unclear exactly how "post-war" norms apply to cyber-war, or broad-based computer attacks. I myself think there's much room for both manoeuvre, and hard, ground-breaking work, on this subject. All I wish to point out is that there *is* a post-cyber-war phase, just as there is a post-conflict phase for every other kind of armed conflict, and so some principles of post-cyber-attack justice must come into play. I myself lean towards the rehabilitative model, more broadly, for reasons I've detailed exhaustively elsewhere (Orend 2012), and so I would insist above all on some kind of norm of potential **"Clean-up, and Aid with Restoration"** following a cyber-strike. Now, obviously, it depends crucially on the details of the strike: Stuxnet, e.g., evaporated and didn't cause spill-over damage to civilians, and so it's hard to see what duties of clean-up might meaningfully have been called for. But in the Russia/Estonia case, where people may have suffered real (mainly financial) hard-ship during their week of being blocked out from their banks, and not having access to government services, etc., some kind of actual monetary restitution might be in order.

Relatedly, it seems that there would be a *jus post bellum* norm calling for **"Public Accountability"**, in terms of a public declaration of why a country resorted to a cyber-strike, and/or why it responded either kinetically or in a cyber way, to a cyber-attack. Both *jus ad bellum* and *jus post bellum* unite together to call, very strongly, for public accountability and transparency both before, and in the aftermath of, war.

As war crimes trials are called for *après la guerre*, so it would seem that cyber criminals need to be held accountable, and investigated for charges, following a cyber-strike. Such **"Trials for Cyber-Criminals"** would serve to underline and enforce the seriousness of their actions, and the attitude of the international community towards things like theft of intellectual property, espionage, and especially harm-causing acts of sabotage. Legal innovations are called for here, in order to bring such into reality (Dinniss 2012; Hyacinthe 2011).

Finally, it would seem as though some **"De-cyber-ization"** might be called for, if we follow the logic of demilitarization post-war. If cyber tools were used in an aggressive attack, then the international community, and especially any victims, are entitled to some reasonable security that they will not be made victim once more, in the near future, to the cyber-schemes of the aggressive power. How, exactly, to go about such stripping or curbing of cyber-power is, of course, beyond the ambit of this paper… and the cyber-skills of its author.

1.3.8 Middle Ground Judgment

Now, this third section started off—a while ago—by saying that endorsement would be made of some kind of middle ground, in this case between realism and just war theory. Obviously, given all the effort just now put into describing the utility and sense of applying just war rules to cyber-warfare, it might be wondered how, exactly, I see a middle ground between just war theory and realism in this regard.

First, it must be noted how much middle ground there is *already* between realism and just war theory: many just war rules make not only moral sense but have clear benefits in terms of realistic self-interest. For example, there is clear over-lap between the just war norm of proportionality and the military maxim of **an economy of force** (i.e., don't use more force than is strictly needed, as resources must be conserved and deployed only when most required). Last resort and probability of success could, straightforwardly, be stated either as moral, or as prudential, maxims of action. And, generally, many realists concur that, given war's huge costs and frightful risks, a rational leader should only contemplate war in response to an obvious and overwhelming danger, such as armed attack by an aggressive invader. Even the norm of discrimination and non-combatant immunity, which otherwise seems saturated with ethical intent, turns out to have potent prudential value as well: one only wants one's military resources, and killing force, to strike at actual sources of harm. Taking out civilians, and civilian targets, almost never directly advances military objectives: far better that one's bullets and bombs take out truly strategic targets that are part of the war machine of the enemy society. Relatedly, one can see how wrapping up a war well, and avoiding the creation of future generations of bitter enemies, can not only serve moral ends but also the long-term national interest of a self-regarding political community.

Secondly, in connection with cyber-war in particular, its very newness calls out for the combined resources of traditions of thought as formidable as realism and just war theory. Indeed, the moment today is arguably much like another moment in modern history: in the mid-1940s, when atomic weapons were just invented. (Here, indeed, *is* a legitimate sense in which reference to WWII is helpful and illustrative, as opposed to being off-key and exaggerated.) Now, as then, there's a brand-new technology of very considerable power and implication. There's absolutely no law regulating its use. Every country thus must do a calculus of self-interest to see how and whether this new tool fits into its self-image, its values, and its overall foreign-policy strategy. This is the least, we might say, that it owes its own people. From there, attempts can then be made to forge the equivalent of arms control agreements, bringing the technology into line and striving to keep it out of the hands of the most dangerous actors.

1.4 Optimism Vs. Pessimism

Which brings us to the final distinction: will we be able to achieve such control, such progressive agreement about when it is proper, and when illegal, to use cyber-warfare? The optimist says: why not? If we did it with something as ferocious as atomic and nuclear weaponry, we can do it with cyber-war technology. The pessimist would be inclined to cite how different cyber-technology is, how widespread and diffuse and more easily hidden it is, and comment darkly as to how, in many ways already, the world has devolved into a situation where, in cyber-terms, it is somewhat like a Hobbesian war of everyman against everyman, or at least every country against every country (Dipert 2011). The middle ground judgment here, in my view, would thus be that, while the pessimist probably provides an accurate description of the state-of-play as it presently stands, there are some historical grounds for believing that, if we've been able to bring other forms of very destructive technology under control through international laws and arms control agreements, then we ought to be able to do the same things with the tools of cyber-war (Ventre 2010, 2011).

1.5 Conclusion

This paper—striving to dispel some of the fuzzy fog surrounding the fifth domain of warfare—first sought to define its terms, and then to consider in a substantial way three "big picture" distinctions surrounding informational warfare: (1) that between cyber-war-skeptic and cyber-war salesman; (2) that between realism and just war theory; and (3) that between optimist and pessimist. With regard to each distinction, it was argued that a middle ground judgment between the two seems the best and most promising way to understand the issue, and to wrestle with the many, and profound, challenges which cyber-war technology poses to the community of nations.

References

Aristotle. 1984. *The politics*. Chicago: University of Chicago Press (Trans. by Carnes Lord.).

Brenner, S. 2009. *Cyber threats: The emerging fault lines of the nation-state*. Oxford: Oxford University Press.

Carr, J. 2010. *Inside cyber warfare*. London: O'Reilly.

CBS News. 2009. *60 min*. broadcast Nov. 06, 2009.

Clark, W. 2002. *Waging modern war*. New York: Public Affairs.

Clarke, R. 2010. *Cyber war*. New York: Harper Collins.

Clarke, R. 2012. *Cyber-war: The next threat to national security and what to do about it*. New York: Ecco.

Clausewitz, K. 1995. *On war*. Harmondsworth: Penguin, with quote at 64 (Trans. by A. Rapoport.).

Cook, M. 2010. Cyberation' and just war doctrine. *Journal of Military Ethics* 2010:417–422.

Dinniss, H. 2012. *Cyber-warfare and the laws of war*. Cambridge: Cambridge University Press.

Dipert, R. 2011. The probable impact of future cyberwarfare, paper delivered at *The First International Workshop on "The Ethics of Informational Warfare"*, University of Hertfordshire (UK), July 1, 2011.

Dobbins, J., and S. Jones, eds. 2007. *The United Nations' role in nation-building*. Washington, DC: RAND.

Dobbins, J., et al. 2003. *America's role in nation-building*. Washington, DC: RAND.

Floridi, L. 2010a. *Information*. Oxford: Oxford University Press.

Floridi, L., ed. 2010b. *The cambridge handbook of information and computer ethics*. Cambridge: Cambridge University Press.

Gross, M. 2011a. The Fog of Cyber-War, *Vanity Fair* (April 2011), 155–198, with quote at 198.

Gross, M. 2011b. Enter the Cyber-Dragon, *Vanity Fair* (Sept. 2011), 220–234.

Hyacinthe, B. 2011. *Cyber-warriors at war*. New York: XLibris.

Ignatieff, M. 2001. *Virtual war: Kosovo and beyond*. London: Picador.

Johnson, J. T., and G. Weigel, eds. 1991. *Just war and gulf war*. Washington: University Press of America.

Kant, I. 1983. *Perpetual peace and other essays*. Indianapolis: Hackett (Trans. by T. Humphrey.).
Karatzogianni, A., ed. 2008. *Cyber-conflict and global politics*. London: Routledge.
Keegan, J. 1990. *The second world war*. New York: Vintage.
Keegan, J. 1994. *The first world war*. New York: Vintage.
Kissinger, H. 1995. *Diplomacy*. New York: Harper Collins.
Lucas, G. 2011. Just War Theory and Cyber-War, paper delivered at *The First International Workshop on "The Ethics of Informational Warfare"*, University of Hertfordshire (UK), July 1, 2011.
Lynn, W. J. 2010. Defending a new domain: The pentagon's cyberstrategy. *Foreign Affairs* 2010 (Sept./Oct.): 97–108.
Machiavelli, N. 1998. *The prince*. New York: Penguin Classics.
Menthe, D. 1998. Jurisdiction in cyberspace. *Michigan Technology Law Review* 69 (1998): 6–52.
Morgenthau, H. 1970. *Politics among nations*. 5th ed. New York: Knopf.
Murray, B., and R. Scales. 2003. *The Iraq war*. Cambridge: Harvard University Press.
New York Times. 2010. Malware Hits Computerized Industrial Equipment. *New York Times*, 24 Sept.
Orend, B. 1999. Crisis in Kosovo: A just use of force? *Politics* 19 (1999): 125–130.
Orend, B. 2000a. *War and international justice: A Kantian perspective*. Waterloo: Wilfrid Laurier University Press.
Orend, B. 2002b. Justice after war. *Ethics and International Affairs* 16 (1): 43–56.
Orend, B. 2006. *The morality of war*. Peterborough: Broadview.
Orend, B. 2009. *On war: A dialogue*. Lanham: Rowman Littlefield.
Orend, B. 2012. Justice after war: Towards a new Geneva convention. In *Ethics beyond war's end*, ed. E. Patterson, 175–196. Washington, DC: Georgetown University Press.
Orend, B. 2013. *Introduction to international studies*. Oxford: Oxford University Press.
Price, M. 2007. *The Wilsonian persuasion in American foreign policy*. New York: Cambria.
Rawls, J. 1993. *Political liberalism*. New York: Columbia University Press.
Regan, R. 1996. *Just war: Principles and cases*. Washington, DC: Catholic University Press of America.
Roberts, A., and R. Guelff, eds. 1999. *Documents on the laws of war*. Oxford: Oxford University Press.
Schmidt, H. 2006. *Patrolling cyberspace*. Washington, DC: Larstan.
Stallwood, O. 2012. Flame virus: How malware became the new weapon of war. *Metro News*, London, UK, 26 June.
The Economist. 2010. Special report on cyberwar: War in the fifth domain. *The Economist* 2010 (July 1): 18–26.
U.S. Congressional House. 2011. *Computer security: Cyber-attacks and war without borders*. Washington, DC: Books LLC.
USJFC. 2010. *Cyberwar report*. released Feb. 18, 2010. www.jfcom.mil.
Ventre, D. 2010. *Cyberguerre*. Paris: Hermes-Lavoisier.
Ventre, D. 2011. *Cyberespace et actueurs du cyberconflict*. Paris: Hermes-Lavoisier.
Walzer, M. 1977. *Just and unjust wars*. New York: Basic Books.
Woodward, B. 2004. *Plan of attack*. New York: Simon and Schuster.

Chapter 2
The Future Impact of a Long Period of Limited Cyberwarfare on the Ethics of Warfare

Randall R. Dipert

Abstract In this essay, I will first summarize some of the main and most controversial published claims in my recent work on ethical considerations in cyberwarfare (*The Ethics of Cyberwarfare*, Journal of Military Ethics). I will then expand and critique some of these claims. Finally, I will turn to discuss some of the ways in which information systems, the internet, and even international relations may change because of a coming era of cyberwarfare.

2.1 Introduction

In this essay, I am going to begin by summarizing some of the main and most controversial published claims in my recent work on ethical considerations in cyberwarfare (Dipert 2010). I would then like to expand and critique some of these claims. Finally I will turn to discuss some of the ways in which information systems, the internet, and even international relations may change because of a coming era of cyberwarfare.

A brief comment. Some of my claims do not fit perfectly well into Just War theory and contemporary accounts of the morality of war because certain claims in game theory and the theory of conflict have important ramifications for moral theory that have been largely ignored (Dipert 2006a, b). One precept of a moral and prudential theory of war for me is that one must consider the likely effect of applying its principles over time—how it is likely to affect all parties' rational behavior. For example, the seemingly sensitive *jus in bello* policy of *never* risking innocent hostages lives is likely, over time, to result in many more hostages being taken and put at risk. This seems to be a stark and powerful counterexample to proposals such as Jeff McMahan's that lives in war can only be taken if the targets have themselves incurred what he calls "liability"—a complicated legal term of art (McMahan 2011). Certainly some human shields have incurred no liability whatsoever.

R. R. Dipert (✉)
Department of Philosophy, University at Buffalo, Buffalo, NY, USA
e-mail: rdipert@buffalo.edu

L. Floridi, M. Taddeo (eds.), *The Ethics of Information Warfare,*
Law, Governance and Technology Series 14, DOI 10.1007/978-3-319-04135-3_2,

Yet a policy of protecting them from admittedly undeserved risk of death is likely to expand the number of future hostages put at risk by an unscrupulous enemy. On the other hand, applying a seemingly harsh and intuitively unjust principle like tit-for-tat is likely to decrease violence over time, particularly if an unprovoked attack on you is answered by a more destructive counterattack.[1] My views mediate between the more common moralities of war and the usually maligned "realist" position in international affairs (Dipert 2006a, 2006b).

2.2 Recent Claims

Now to my past claims:

1. Terminology and Taxonomy. It is important to separate the notions of cyberwarfare between nations or nation-like political entities from cyberattacks by individuals, corporate entities, or other groups of individuals, and to separate these from cyberespionage and from cybertheft of intellectual property. Likewise, it is important clearly to separate defensive cyberwarfare (cybersecurity) from offensive cyberwarfare, including when offensive cyberwarfare is ultimately "defensive" in nature, used or threatened as a deterrent.
2. I have argued that the Attribution Problem does not prevent morally justified cyber counterattacks, as some have suggested. It is true that there is now, and will likely remain in many cases, an Attribution Problem—uncertainty about the source or intent of a act of cyber warfare—but this need not absolutely prevent counterattack. This especially affects internet-based cyberwarfare but also other forms. As with other forms of uncertainty (such as in preemptive or preventive war) a morally permissible counterattack may require a threshold of likelihood that is tied to the risk of responding, and not responding. The issues turn on little-discussed moral ramifications of epistemic dimensions of attack and counterattack: to what degree we know that an enemy will, or has, attacked us (as well as who this enemy is, and their intent).
3. Counterattacks. Justified response to a cyber attack may be a conventional or a cyber counterattack. If a cyberattack should result in the death of human beings or in destroying physical objects, then this may justify a cyber counterattack or even a conventional counterattack. This doctrine has since appeared as a part of the emerging U.S. Cyber Defense Policy. In a wise policy, the exact thresholds for responding to a cyberattack, what kinds and with what level of force, should remain secret.
4. The Ontological Problem. Although conventional harm by acts of war, such as causing deaths or physical destruction, is—regardless of weapons—undoubtedly

[1] The positive characteristics of such retaliatory strategies of conflict were popularized in studies by Robert Axelrod and others, but the main outline of the view, including its application to international affairs, had long before been described by Thomas Schelling and others (Schelling 1960, 1980).

covered by existing international law, treaties, and by most traditional moral theories of war, the most likely harms by acts of cyberwarfare are not. Attacks in cyberwarfare often will inflict a distinctive form of harm: they may not kill or wound human beings, or even destroy or damage the physical objects of value, but may impair the *functioning* of systems that are important for welfare and happiness, such as systems of making economic transactions or of communication. This harm may be severe in no single region or facet of a culture, but may be a highly distributed harm, afflicting tens or hundreds of millions of people in important, but non-life-threatening ways. This problem is thus twofold. The kind of harm is not one dealt with in traditional moral and legal theory of war, and there is a problem of the possibly large *sum* of harms. It would nevertheless appear to be part of morally justified national defense to limit, or punish, such an intentionally inflicted sum of harms.

Again the problem is with policies, and over the long term: always to ignore such attacks will encourage future such attacks and damage, up to the level that traditionally has justified serious counterattack, such as the "armed attack" condition of Article 51 of the U.N. Charter. Perhaps the most widespread view about cyberwarfare in international law is that of Michael Schmitt, who holds that only cyberattacks rising to the level traditional acts of war (including permanent physical damage or deaths) permit a lawful response without Security Council approval (Schmitt 2002; Schmitt et al. 2004). This points to a careful legal distinction, between acts that are not sanctioned or deemed lawful, and those that are unlawful, i.e. contrary to international law. Without some clear answers to this question, traditional understanding of international law is mute about the most common likely acts of cyberwarfare.

5. "Black hat hackers in the basement." The ubiquity of the necessary tools (computer, internet connection, thumb drive,…) and of the skill necessary to produce a cyber attack (DDoS or creating and distributing malware) makes treaties limiting cyber attacks unfeasible—at least for now.[2] This contrasts with the situation of NBC weapons, with their exotic ingredients and devices to deploy them, with the skills necessary to produce or weaponize them, and even with the production and distribution of larger or specialized military weapons, such as artillery. Consequently a treaty limiting the production or application of cyberweapons would be like a treaty controlling the production and application of knives or "harmful information."
6. Almost all of the literature simply assumes that attacks via the internet are the only or at least dominant form of cyberwarfare. In fact there have been only four well-known acts of cyberwarfare. These are the Russian "patriot hacker attacks" on Estonia—an indiscriminant attack that interfered with the functioning of civilian information infrastructure (and was possibly not initiated and controlled by the Russian government), the Russian attacks on Georgia government and military information systems in 2009 in conjunction with a short conventional

[2] Even cyberwarfare treaty optimists suggest it may be decades until we have the verification tools and other agreements necessary for meaningful verification. See (Rowe et al. 2011).

war, the 2007 Israeli disabling of parts of the information system for Syrian air defenses in their attack on a nuclear reactor, and the 2010 Stuxnet attack on Iranian uranium processing facilities (Libicki 2012) and (Clarke and Knacke 2010). The first two were primitive DDoS attacks, which technologically sophisticated nations can now avoid or mitigate. The last two involved intentional and permanent damage to physical structures, and thus could have justified some military response by the traditional laws and ethics of military conflict. However these same last two attacks involved means of delivery of malware that was *other than via the internet*. I shall call these OTI (other than internet) acts of cyberwarfare (Dipert 2013a). The Israeli attack on Syrian air defenses apparently involved the physical manipulation of optical cables on Syrian territory or the previous insertion of an integrated circuit (chip) with a backdoor, both technically difficult feats. The Stuxnet attack involved a thumb drive (or perhaps some other storage medium). These two attacks show how shortsighted the preoccupation with the internet have been, and how broad and diverse are the means by which malware may be injected into an information system or by which transmitted data can be altered. This includes embedding of devices containing information-altering software in any hardware, including peripherals and modularized devices (such as disk drives), bluetooth-enabled devices, the embedding of radio-transmitting and receiving chips in hardware, perhaps communicating at very low power levels and using unusual radio-wave conduits, such as the building wiring. As the bugging of the U.S. Embassy in Moscow during the Cold War shows, there are remarkable and nearly undetectable possibilities for forwarding (and injecting) information.

OTI acts of cyberwarfare (and cyberespionage) will have some of the same characteristics of internet cyberwarfare, such as their own attribution problems. After all, chips could be surrepticiously placed, unwittingly or not, by any person who has physical access to any component of an information system at any point in its manufacture, assembly, and transportation. So "air gapping" is only a marginal, short-term solution to internet attacks, and even if techniques and agreements could remedy the internet Attribution Problem, they offer no defense against myriad other techniques of interfering with the functioning of information systems.

7. Defensive cybersecurity as the sole or main national defense strategy is closely analogous to depending only on a fixed defensive line (e.g., the Maginot Line), or to to a nation only building better bomb shelters in the era of bombers and nuclear weapons (or for Israeli policy, developing better warning systems, better and more shelters for long term inhabitation, and a hugely expensive system like Iron Dome against Katyusha and Grad rockets). Ultimately, it is folly: offensive weapons always have a tactical and financial advantage over purely defensive countermeasures.

   A purely defensive cyberwarfare policy creates a disproportionate burden on the attacked nation rather than the attacking nation. Provoking expensive countermeasures is itself an act of economic warfare. It is almost certain that some nations (in part hiding behind the Attribution Problem) will attack others. Protective measures by the U.S. government currently are estimated to be in the 10's of billions of dollars and are widely expected to climb into the 100's of billions within years.

8. Cyber Cold War. Because some nations and political organizations are technologically sophisticated and likely to attack others for political and economic reasons, and because of the difficulties posed by the previous considerations, we are likely to experience for some decades multilateral acts of cyberwarfare and counterattacks. We are likely to experience a prolonged Cyber Cold War dominated by skirmishing and deterrent counterattacks (as well as unprecedentedly aggressive espionage and theft–although with differences from the actual Cold War, which was mainly conditioned by the possession and testing but not use of nuclear weapons. These multilateral attacks will largely be between technologically advanced nations, without regard to region of the world and perhaps only weakly limited by traditional alliances and friendships.

2.3 Comments on these Points

1. Taxonomy and Terminology. One aspect of cyberwarfare (and the injection of malware and impairing of functioning of information systems) that is still not much discussed consists in attacking information systems through *other* than the internet. These are OTI attacks and intrusions, described above. Information systems can be attacked or corrupted by a wide array of software and hardware points of entry. There is also the danger that malware could be embedded in distributed operating systems and large applications (e.g. Adobe Acrobat), particularly when access to source code is not public. Another danger lurks in the design and manufacture of hardware that may be manufactured outside of a country. Malware and espionage circuitry can be easily embedded in hardware and firmware. This extension of "cyber" beyond internet activities is little discussed but has been noted by Michael Chertoff, former U.S. Secretary of Homeland Defense, who calls it a "supply chain problem." We are likely also to see the development of devices that can inject at a distance malware and faulty data into, or disturb the functioning of information systems through, unshielded electrical pathways without utilizing the internet. Because they do no generate detectible, external magnetic or other fields, and because they are much more secure to intrusion, optical networks and eventually optical computers will significantly lessen these dangers.[3]

[3] Dual wires were used extensively for military telegraphy of secret information as early as the U.S. Civil War (including using rails) and for telephony in WW I. In that latter war it occurred to engineers that it would vastly simplify the stringing of wires if one used just one wire, using the ground (the earth itself) to complete the circuit. However German engineers realized this, and by tapping the earth at two places between the ends of the circuit, were able to detect the minute currents and effectively read the messages. It is now possible to detect magnetic and radio fluctuations at truly minute levels (as utilized in MRI technology), and, because the currents are so slight in

So by "cyberwarfare" I include non-internet attacks on information systems. We need to think out-of-the-box of narrowly *internet* cyberwarfare, to a broader notion of warfare involving information systems of all sorts, including diverse forms of psychological operations, cyberespionage, and disinformation campaigns. The wider role of information, and information warfare, is developed in other papers in this volume.

One area in which my earlier distinctions were not sufficiently articulated concerns what counts as a state or political organization. I extended cyberwarfare to include acts by state-like political organizations. This change is necessary because of the changing nature of what is considered conventional warfare, and follows the lead in Brian Orend's definition of war in his excellent article in the *Stanford Encyclopedia of Philosophy* (Orend 2005). However there has been further development in the discussion in the increasingly sophisticated literature on forms of cyberattack. One particular development that has been noted is the formation of, and very effective cyberattacks by, loose federations of black-hat hackers.

Cyberespionage or cybertheft (of intellectual property) can be committed by individuals, groups of individuals, with or without state support or direction. One of the most common forms of this is when the attackers seek to benefit financially from their actions. However, what would normally be cybercrime even if performed by a state shades into political action, and thus cyberwarfare, if the attacking nation seeks not just financially to itself profit but primarily to damage another state or its economy.

A recent distinction divides organized cyberattacks into three categories: cybercrime, cyberwarfare, and hacktivism. Criminal cyberattacks are those motivated primarily for their own financial gain, and would include most theft of credit card numbers, passwords, and so on. Cyberwarfare cyberattacks are attacks on one state primarily on another state. Hacktivist cyberattacks are sometimes mere vandalism and sometimes conducted for some political. The massive Wikileaks publication of classified U.S. documents, attacks by Anonymous, and the LulzSec attacks on various government agencies in the U.S. and the U.K., are of this sort. It becomes difficult in practice sharply to distinguish a hacktivist attack that is in support of a vague doctrine of anarchy or anti-authoritarianism, from disorganized black-hat hacking attacks, in which the hacker may do it mainly for the thrill of notoriety or as a display of technical expertise. (Political hacktivist attacks could also be described as acts of terrorism, at least if the intentional targets are civilian.)

Within cyberwarfare attacks, there at least three kinds: (A) those ordered or directed by a state's central authority, i.e., acts of war. Call them Commanded. (B) those that benefit the host state, are not under traditional military command and control, but are knowingly tolerated by the host state, call them Tolerated or Harbored, and (C) a third category of cyberattacks on behalf of a state but that are not expressly tolerated by the benefited state, perhaps because it does not know of them. Call them Patriotic.

---

modern digital wires, it is also possible to disrupt them using an external device if they are unshielded or poorly shielded. (And, in the worst case of EMP weapons, to fuse circuitry and destroy the semiconductors.)

Lacking intelligence about the exact constitution of the group involved in organized attacks against Gmail from a given IP address and building in China, and the hackers' motives, China has sought to cloak itself from liability by arguing that these were—for all Google and the U.S. knows—of the Patriotic type (C) that have no direct relation to the command and control of the central Chinese government. Since the Chinese did not volunteer verifiable information about how they would stop this group from continuing their attacks, the attacks were more likely Tolerated attacks (B) or even secretly Commanded attacks of type (A) I think the distinctions among cyberattacks related to a central government or authority will become of greater and greater importance. The existence of Patriotic (C) attacks in the industrialized democracies, together with the Attribution Problem and in particular an enemy's inability to distinguish these from type Commanded or Tolerate attacks for which the state *is* responsible, will likely require monitoring by governments of hackers within their territory who are damaging other countries' governmental and economic institutions. In the U.S. legal context this would require application of the rarely-invoked Neutrality Law (of 1794, with many alterations and now in U.S. Code Title 18 I.45 par 960) that forbids attacks on other nations by private American citizens. I assume other nations have similar laws—or should have them. The phenomenon may also bring about future measures that infringe on real or supposed information rights.

A full table of distinctions looks like this:

1. Cybercriminality (for profit).
   1.a. By individuals
   1.b. By organizations (hacker mafias)
   1.c. By states or political organizations (where their motivation is not primarily political).
2. Hacktivism (including clearly anarchic or anti-authoritarian black-hat hacking)
   2.a. Individual hacktivism
   2.b. Organized hacktivism
3. Cyberwarfare: attacks on one state primarily for the benefit of another state or political organization
   3.a. Initiated or directed by a state's command-and-control apparatus
   3.b. Not initiated and anddirected cyberwarfare attacks, but tolerated by the host state
   3.c. Cyberattacks on behalf of a state against another state, but neither directed by nor expressly tolerated by the host state.

2. The Attribution Problem. The Attribution Problem appears, I would now argue, to be much less of a problem than it did even a year ago. For one thing, forensic technology has advanced rapidly (although many of its techniques remain secret). It was once a serious problem to locate the original source of botnet attacks and thus difficult to deter the creators of botnets. However, by saving not just present but historical data of IP packets passing through various servers, nodes, and exchange points, and with the devising of software to sift through this massive amount of

data, the first instances of a botnet's software can often be identified. Other forensic technologies have developed apace with these, such as by scrutiny of code by large teams of the most sophisticated cybersecurity experts. As an example, see the Stuxnet White Paper produced by the cybersecurity firm Symantec.

In support of rejecting absolute, objective certainty about the attacker and motive as *necessary* for just war, consider a more mundane application of principles of *jus in bello*. A commander believes, based on considerable evidence, that an attack on the commander's soldiers came from a certain building. The commander has taken all reasonable measures to acquire more information about the likely presence of enemy soldiers as well as the likely absence of civilians. Despite a lack of certainty, it seems to me, using traditional reasoning about morality in warfare, the commander is morally justified in destroying the building if the overall military objective is sufficiently important. This is a *jus in bello* case of uncertainty.

One might argue that the evidence morally required to initiate a war is much higher for *jus ad bellum* cases than in *jus in bello* cases. However, I think it is the case that we are already in the "in bello" situation. We are not sure in real time of who launched a given cyberattack on us, but we are highly justified in believing that certain nations have launched attacks (or tolerated or harbored the attackers).

3. Against Cyberwar Treaties' Effectiveness. This is more complex point than I first thought it was. My original argument against the desirability of treaties was that effective treaties require verification, and that neither the equipment nor skills necessary for cyberwarfare are detectable. There are at least three ways in which treaties against the development and use of cyberweapons might proceed. First, since forensic techniques, especially for identifying a source IP address, are rapidly improving, then if there were a treaty that allowed international investigators unfettered access to all locations, investigator could be sent to those sites to investigate the site, its computers, and the service providers to which it as connected. Such procedures have at least been attempted in the case of suspected NBC weapons of smaller powers. However, large powers are not likely to accede to this verification, and if they do, they will probably often thwart the efforts. Second, traditional forms of intelligence, both human and signals intelligence, might also serve to locate probable cyberweapon development centers and cyberwarrior training sites. If one assumes that the development of highly sophisticated cyberweapons requires the cooperation of even platoon or company-size units, this would also help.

Finally, if every country had laws criminalizing unauthorized attacks on foreign nations (like the U.S.'s Neutrality Act), and if extradition treaties were extended to cover these crimes, then this would at least allow the suppression of a host nation cyberattacks that were merely tolerated (type (B) attacks). However mutual extradition treaties do not even exist in a number of cases (such as of the U.S. with the Russian Federation, with the Ukraine, with China, and with rogue nations such as Iran and North Korea). Even in the tawdry case Dominique Strauss-Kahn, NYC Police had the reasonable concern that he could not be extradited from France, ostensibly an ally, for a sexual crime. The task of negotiating cybercrime extradition treties seems hopeless. Furthermore, extradition treaties typically list crimes to which they apply, mainly violent felonies. They often explicitly exclude military and political crimes.

The main thread of my argument is to argue that international damage from cyberattacks can only be mitigated in the future by deterrent strategies. The argument in favor of deterrent strategies also requires that there be some likelihood that deterrence *can* work to limit the development of and use of cyberweapons. Things are not so simple however. As various authors have pointed out (Libicki 2009) there are a number of preconditions for the success of deterrent strategies. It requires that a nation highly values something that other nations can threaten. It also requires some limited game-theoretic rationality on behalf. These conditions appear to have been met in the Cold War, but a related argument concludes that something like the harboring doctrine should be employed. This doctrine is a part of what sometimes called the George W. Bush Doctrine, from his statement of it after shortly after the attacks of September 11, 2001. The full doctrine was that the U.S. will see no difference between nations that attack us and nations that harbor those who attack us.

2.3.1 Future Cyberwar Defense

One way to proceed in the future is simply to continue with current techniques and do a better job using them. This is essentially having large cybersecurity organizations, like Symantec and MacAfee, disseminate information and devise protection. Government agencies and large corporations, especially in the financial defense sectors, will have their own experts and alternative or additional software beyond what is publicly available. End-users may be involved since one may increase layers of passwords and challenge questions, add captcha tests, and requiring frequently changed passwords. Communication between users should encrypted, with increasingly sophisticated methods.[4]

Still other techniques, in order to gain access to a system or to files through the internet or any connection, are to require specialized software or hardware (The U.S. DoD's CAC card and reader) to be installed on a computer, or to require extensive and complicated handshakes before data can be exchanged.

Another but very expensive solution would be to have a secure network running only proprietary pieces of software specially written for computers attached to it, that are comparatively small, and clean. At the extreme end of this technique one uses a proprietary operating system. They might have many "write" functions disabled. One would give up on large commercial applications, and instead use modified open source software and an operating system such as a modified Linux, whose every line of code has been examined and understood by an agency's experts.

There are at least two reasons why these "traditional" cybersecurity techniques, even altogether, are likely to be inadequate. First, no matter how sophisticated these protections at first seem, all of such techniques have turned out to be penetrable by

[4] One little noted possibility is that even the most sophisticated encryption techniques become quickly breakable, either by hardware techniques or algorithms. A worse possibility is that the widespread assumption that, in computational complexity theory, $P \neq NP$. If this turns out incorrect, all encryption techniques are breakable.

sophisticated and persistent hacking. Unfortunately, high-profile supposedly secure sites have been virtual red flags for the most accomplished and persistent attacks. (Perhaps one could hide the sites themselves, change IP addresses like frequency shifting radio communications—with hundreds of decoy sites with sophisticated barriers. Such possibilities are made more feasible by the vast increase in IP addresses with the shift from IPv4 to IPv6).

The second reason is that these techniques ultimately shift the cost to defenders and are thus variations on the build-a-better-bomb-shelter theme. Such static hunkering down has proven to be poor technique in conventional warfare.

Ultimately then, in the long run, the best cybersecurity is likely to be provided by deterrent or punitive strategies. Even these will have limits, such as with undeterrable attackers.

One approach, which some have forecast is the best option, is to envision large consortia of private and government cybersecurity systems, sharing information in real time, and rapidly identifying common threats and sharing defensive techniques. Various forms of such information swapping markets and mandates were proposed in the various cybersecurity bills in the US in the first half of 2012. There are at least three reasons why this approach cannot be expected to be very effective soon or in the long run. One is that the identification of threats and attacks, and their categorization, requires huge amounts of information to be transmitted and immediately understood by both by systems using greatly different ways of categorizing these threats. This information, or "incident reports", would include categorizations of the unusualness of the emails, IP addresses, malware signatures, a description of the vulnerability it appears to exploits (which may be as obscure as a driver, printer spooler, document or video file), its end motive, apparent source and time it first appeared. About one famous virus, Conficker, its end-goal is still not known. Such a distribution system for incident reports, and for remedies, would require a common ontology (a lexicon) that categorizes these diverse aspects, and a clearinghouse, that are both a long way in the future—perhaps a decade or more (Dipert 2013b).

Second, the more detailed and specific such information is, the more likely it is that it will generate a disproportionate number of false alarms. It is also likely that considerable damage may be done—including to malware-detection software—before it can be detected by even the most responsive early warning system. (These are 0-day malware attacks.)

Third, one of the most serious weakness to such a cyberdefensive system would be institutional-political. Consider an attack on a highly sensitive system, such a computer system of the US CIA or Department of Defense. There could be considerable institutional incentive not to alert other potential victims, and instead hide the nature of the attack, its damage, and possible ways to defend against it. For by doing so, an agency encourages an enemy into believing that it has been undetected and has possibly succeeded. If an agency leaks into the general cybersecurity community information about the threat, its degree and form of penetration, and damage, and possible antidotes, one essentially hands the enemy a detailed After Action Report. It is a well-known technique to make an enemy believe its espionage techniques or weapons have been successful. In military terms one of the most dangerous forms of information one can feed an enemy is precisely weapons and modes of attack have been successful or not.

Consequently, it is fairly likely that an attacked agency would not wish to disseminate its vulnerabilities, the success of attacks on it, and defensive measures into the open general public (or even into the community of cybersecurity experts). On the other hand, government agencies could compel with the force of law individual citizens and corporations to share with them any reasonable suspicions of malware. The problem of data interoperability (common ontology) and the quantity and poor quality of such suspicions, would remain. Cyberincident reporting would be like having a criminal hotline in which anyone, in any language and with various degrees of observational astuteness, would report any suspicions about crimes they might have witnessed, including traffic violations.

One proposal I made in an earlier paper, to which I see no technical, legal, or moral objection is to make Internet Service Providers (ISPs, including at the higher tiers of internet traffic routing) responsible for the behavior of computers that attach to their network. It is only through this medium that individual users have access to the internet. By blocking users whose computers have not been thoroughly checked by the ISP itself (as part of a handshake when first communicating with the IP), such an arrangement could at least block the most persistent virus and botnet problems. It is also at this level that IP address masking (spoofing) in all communications, including email, would be most easily detectable.

ISPs themselves could be punished for various degrees of non-compliance, from a fine for failing to check attached computers for viruses and botnets, up to disconnection from the internet for the service itself, and all of its users. This might even apply to an entire country's domain.

Because of its highly distributed nature, and because of a philosophy of open access, there is no authority for the internet as a whole, except perhaps for the Internet Corporation for Assigned Names and Numbers (ICANN). It has restricted its actions to purely technical ones (such as the switch from IPv4 to IPv6). It is nevertheless conceivable that it could be turned into a weak internet policing system, by invalidating an IP address, or a group of IP addresses due to some pattern of misuse of the internet. A more likely scenario would be within one legal jurisdiction and would regulate the addresses an ISP may communicate with, and especially at the internet exchange points and information conduits within that territory or under the control of that territory (e.g., satellites). Implementing any such scheme would be nearly impossible in the present atmosphere of glorification of the freedom of the internet. However, I do not know of any sophisticated political philosophy that would grant the privacy and anonymity rights, and complete freedom of communication, that the dominant internet philosophy appears to advocate. If anything, this idealized internet is a nearly perfect implementation of the Ring of Gyges (in Plato). A far more helpful attitude toward information would be to regard some information as poisoned: it does intended harm to its users.

However even such a complete scheme of legal liability for ISPs or regulation of the internet would not limit the dangers of first-use (zero-day) internet weapons. While there may eventually be sophisticated software that identifies data that is likely to be malware, it would probably admit some malware and block some beneficial data. Furthermore, such controls would only block internet-borne attacks and not those arising from other OTI vectors of information exchange, such as hardware, thumb drives, and CDs and DVDs.

The future cyberworld I envisage is not a neat or a pretty one. It is beset by problems ontological, moral, and computational. But is it necessarily worse than the total destructiveness of the twentieth and early twenty-first century? I don't think so. In fact it might be one in which informational proxy wars are fought, but that humans, other creatures, and the environment are not permanently damaged. It is possible, after all, that the future will *not* be worse than the past.

Bibliography

Axelrod, R. 1984. *The Evolution of Cooperation*. New York: Basic Books.

Christopher, P. 1999. *The ethics of war and peace: An introduction to legal and moral issues*. 2nd ed. Upper Saddle River, NJ: Prentice Hall.

Clarke, Richard, R. Knake. 2010. *Cyber war: The next threat to national security and what to do about it*. New York: HarperCollins.

Dipert, R. R. 2006a. Strategies, rationality, and game theory in the philosophy of war, paper. Joint Service Academy Conference on Professional Ethics (JSCOPE now ISME). Springfield, VA. http://isme.tamu.edu/JSCOPE06/Dipert06.html. Accessed 3 Nov 2013.

Dipert, R. R. 2006b. Preventive war and the epistemological dimension of the morality of war. *Journal of Military Ethics* 5 (1): 32–54.

Dipert, R.R. 2010. The ethics of cyberwarfare. *Journal of Military Ethics* 9 (4): 384–410.

Dipert, R.R. 2013a. Other-than-Internet (OTI) warfare: Challenges for ethics, law, and policy. *Journal of Military Ethics* 12 (1): 34–53.

Dipert, R.R. 2013b. The Essential Features of an Ontology for Cyberwarfare. In *Conflict and Cooperation in Cyberspace,* ed. P. Yannakogeorgos and A. Lowther, 35–48. New York: Taylor & Francis.

Falliere, Nicholas, L. O. Murchu, Eric Chien. 2011. Symantec Corporation: W32.Stuxnet Dossier. www.symantec.com/content/en/…/whitepapers/w32_stuxnet_dossier.pdf. Accessed 1 Feb 2011.

Libicki, M. 2009. *Cyberdeterrence and cyberwar*. Santa Monica: RAND Corporation.

Libicki, M. 2012. Panel on response to cyberattacks: The attribution problem. The McCain Conference, organized by the Stockdale Center for Ethical Leadership. U.S. Naval Academy, Annapolis MD April, 2012. http://www.youtube.com/watch?v=bI7TLqTt0H0. Accessed 3 November 2013.

McMahan, Jeff. 2011. *Killing in war*. Oxford: Oxford University Press.

Orend, B. 2005. War. In *The Stanford Internet Encyclopedia of Philosophy*. http://plato.stanford.edu/entries/war/. Accessed 13 Sep 2012.

Owens, W., K. Dam, H. Lin. 2009. *Technology, law, and ethics regarding US acquisition of cyberattack capabilities*. Washington, DC: National Research Council of the National Academies of Science.

Reichberg, G., H. Syse, E. Begby. 2006. *The ethics of war: Classic and contemporary readings*. Oxford: Blackwell.

Rowe, N. 2006. A Taxonomy of Deception in Cyberspace. International Conference on Information Warfare and Security. Princess Anne, MD.

Rowe, N. 2009. The ethics of cyberweapons. *International Journal of Cyberethics* 1 (1): 20–31.

Rowe, N. 2010. Toward reversible cyberattacks. In *Warfare and cyber terrorism,* ed. L. Janczewski and A. Colarik. Hershey: Information Science Reference.

Rowe, J., E.J. Crusty. 2010. Deception in cyber attacks. In *Warfare and cyber terrorism,* ed. L. Janczewski and A. Colarik. Hershey: Information Science Reference.

Rowe, Neil, et al. 2011. Challenges in monitoring cyberarms compliance. *International Journal of Cyberwarfare* 1 (1): 1–14.

Schelling, T. 1960, 1980. *The strategy of conflict.* Cambridge: Harvard University Press.

Schmitt, Michael. 2002. Wired warfare: Computer network attack and *jus in bello. International Review of the Red Cross* 84 (8): 346–365.

Schmitt, Michael, H.H. Dinniss, T.C. Wingfield. 2004. "Computers and War: The Legal Battlespace," publication of the International Humanitarian Law Research Initiative, Program on Humanitarian Policy and Conflict Research at Harvard University www.hpcrresearch.org/sites/default/files/publications/schmittetal.pdf. Accessed 5 July 2012.

Walzer, M. 2006. *Just and unjust wars.* 4th ed. New York: Basic Books.

Chapter 3
Is Warfare the Right Frame for the Cyber Debate?

Patrick Lin, Fritz Allhoff and Keith Abney

Abstract Nation-states are struggling to formulate cyberpolicy, especially against foreign-based intrusions and attacks on domestic computer systems. These incidents are often framed in the context of cyberwarfare, which naturally implies that military organizations should respond to these incidents. This chapter will discuss why cyberwarfare is ethically difficult and why, until responsible cyberpolicy is developed, we may plausibly reframe the problem not as warfare but as private defense, i.e., self-defense by private parties, especially commercial companies, as distinct from a nation-state's right to self-defense. The distinction between private defense and national defense is relevant, since victims of cyberattacks have been primarily industry targets and not so much government targets, at least with respect to measurable harm. And we focus on foreign-based cyberattacks since, unlike domestic-based attacks that are usually considered to be mere crimes and therefore a matter for domestic law enforcement, foreign-based attacks tend to raise special alarms and panic about more sinister motives. More than a mere criminal act, a foreign cyberattack is often perceived as an aggression so serious that it may plausibly count as an act of war, or *casus belli*, and so we are quick to invoke national security. But insofar as the state is currently not protecting industry from such cyberattacks—in part because it is difficult to arrive at a sound cyberpolicy—we should consider interim solutions outside the military framework.

Nation-states are struggling to formulate cyberpolicy, especially against foreign-based intrusions and attacks on domestic computer systems. These incidents are

P. Lin (✉) · K. Abney
Department of Philosophy, California Polytechnic State University, San Luis Obispo, USA
e-mail: palin@calpoly.edu

K. Abney
e-mail: kabney@calpoly.edu

F. Allhoff
Department of Philosophy, Western Michigan University, Kalamazoo, USA
e-mail: fritz.allhoff@wmich.edu

L. Floridi, M. Taddeo (eds.), *The Ethics of Information Warfare*,
Law, Governance and Technology Series 14, DOI 10.1007/978-3-319-04135-3_3,

often framed in the context of cyberwarfare, which naturally implies that military organizations should respond to these incidents (Arquilla 2012). This chapter will discuss why cyberwarfare is ethically difficult and why, until responsible cyberpolicy is developed, we may plausibly reframe the problem not as warfare but as private defense, i.e., self-defense by private parties, especially commercial companies, as distinct from a nation-state's right to self-defense.

The distinction between private defense and national defense is relevant, since victims of cyberattacks have been primarily industry targets and not so much government targets, at least with respect to measurable harm (Clarke 2010, esp. Chap. 3; Riley and Walcott 2011). And we focus on foreign-based cyberattacks since, unlike domestic-based attacks that are usually considered to be mere crimes and therefore a matter for domestic law enforcement, foreign-based attacks tend to raise special alarms and panic about more sinister motives. More than a mere criminal act, a foreign cyberattack is often perceived as an aggression so serious that it may plausibly count as an act of war, or *casus belli*, and so we are quick to invoke national security (Gorman and Barnes 2011). But insofar as the state is currently not protecting industry from such cyberattacks—in part because it is difficult to arrive at a sound cyberpolicy—we should consider interim solutions outside the military framework.

In this chapter, though we speak primarily from the US perspective as the one with which we are most familiar, the discussion can apply to cyberpolicies in other nation-states. Further, the issues we identify and discuss are not meant to be exhaustive but only a *prima facie* case for thinking about cyberattacks in a nonmilitary framework.

3.1 Cyberpolicy and Just-War Theory (Lin et al. 2012)

Why it is so difficult to develop responsible policy for cyberwarfare? If we understand war as "actual, intentional, and widespread armed conflict between political communities" (Orend 2005), it is first unclear that a cyberincident is an "attack" or even "armed" conflict. And even if they are acts of war, cyberattacks and counterattacks must adhere to international humanitarian law (IHL), otherwise known as the laws of war. These laws include the Geneva and Hague Conventions, as well as many other international agreements. Much of IHL is rooted in just-war theory, the philosophical tradition meant to establish the moral boundaries of warfare (Aquinas 1948; Walzer 2006; Reichberg et al. 2006). As a general discussion about the ethics of cyberwarfare, let us explain why cyberpolicy is so difficult to reconcile with just-war theory on at least the following five points:

3.1.1 Aggression

By the laws of war, there is historically only one "just cause" for war, a defense to aggression (Walzer 2006, esp. pt. 2). But it is not clear at *what kinds* of cyberincidents are so aggressive that they may be considered to be attacks (never mind "armed" attacks), as opposed to espionage or vandalism. Traditional just-war theory doesn't consider mere (non-military) property damage as *casus belli;* to count as warlike aggression, the act needs to be more serious, such as an actual loss of lives or serious threat of economic harm, e.g., blockade of a trading route (Walzer 2006, Chap. 10). So, on the face of it, taking down a website does not seem to be *casus belli*, to the extent that it is only damage to property.

But in the Digital World, intellectual property is the coin of the realm. A cyber-attack that erased financial data could wipe out entire bank accounts, leaving their owners penniless; this seems to be as severe as a naval blockade. And while the cyber domain (not counting physical substrate, e.g., routers and servers) is composed of only information bits, some of these bits control real-world property, e.g., power grids, nuclear centrifuges, and so on. Therefore, corrupting information data could lead to physical harms. It is a more complicated question, then, whether or not theft of intellectual property, or damage to virtual property, should fall under the threshold for war. Again, it may make a difference as to whether a military website is defaced, as opposed to a commercial website.

Complicating matters further, it is unclear *when* a cyberincident becomes an attack, even if we agree that it is an attack. Is it *casus belli* to install malicious software on an adversary's computer systems but not yet activate it? Or maybe the act of installing malicious software is an attack itself, much like installing a land mine? What about unsuccessful attempts to install malicious software? Do these scenarios count as war-triggering aggression—or are they mere crimes, which do not fall under the laws of war? These questions feature in debates over the legitimacy of preemptive and preventative war (Dipert 2006; Willson 2010).

Another question: Insofar as most cyberattacks do not directly target lives, are they as serious? The organized vandalism of cyberattacks could be serious if it prevents a society from meeting basic human needs like providing food or power, and so could indirectly cause death and injury. A lesser but still serious case was the denial-of-service cyberattacks on media websites in the country of Georgia in 2008, which prevented the government from communicating with its citizens (Markoff 2008). However, the traditional understanding of aggression in just-war theory says that human lives must be directly in jeopardy. This makes it difficult to justify going to war in response to a cyberattack.

3.1.2 Discrimination

The laws of war mandate that noncombatants be avoided in attacks, since they do not pose a military threat (McMahan 2009). Most theorists accept some version of a double effect in which some noncombatants could be unintentionally harmed as "collateral damage" in pursuing important military objectives (Aquinas 1948), though some have more stringent requirements (Walzer 2006). Some challenge whether noncombatant immunity is really a preeminent value (Allhoff 2012), but the issue undoubtedly has taken center stage in just-war theory and therefore the laws of war.

For the military, cyber-counterattacks (or, euphemistically, "active defense") must comply with the principle of discrimination or distinction. But it is unclear how discriminatory cyberwarfare can be: If victims use fixed Internet addresses for their key infrastructure systems, and these could be found by an adversary, then they could be targeted precisely—but victims are unlikely to be so cooperative. Therefore, effective cyberattacks need to search for targets and spread the attack; yet, as with viruses, this risks involving noncombatants.

For instance, consider the uncontrolled propagation of a computer worm such as Stuxnet (Schneier 2010; Sanger 2012). Stuxnet's designers had taken pains in designing it to target only Iranian nuclear processing facilities, yet it had spread far beyond intended targets. If the US is behind Stuxnet, then its own weapon has boomeranged back to US computer systems. Although its damage was highly constrained, Stuxnet's quick broad infection was noticed and required upgrades to antivirus software worldwide, incurring a cost to everyone. The worm also provided excellent ideas for new exploits that are already being used, another cost to everyone. Arguably, then, Stuxnet did incur some collateral damage.

Cyberattackers could presumably appeal to the doctrine of double effect, arguing that effects on noncombatants would be foreseen but unintended. This may not be plausible, given how precise computers can be when we want them to be. Alternatively, cyberattackers could argue that their attacks were not directly against noncombatants but against infrastructure. However, attacking a human body's immune system as the AIDS virus does can be worse than causing bodily harm directly. Details matter; for instance, if it knocks out electricity and the refrigeration that is necessary for the protection of the food supply, starvation could ensue from a modest cyberattack. Disrupting other crucial services, such as hospitals, could also result in deaths, as well as the foreseeable social unrest that routinely accompanies widespread power outages in urban areas.

A serious unintended effect to consider is that any cyberattack or counterattack may need to involve one's own civilian infrastructure (e.g., routers). This is problematic, because in providing material assistance for an attack, the civilian assets involved then can be marked by adversaries as a legitimate target of attack, either cyber or kinetic. For instance, if a counterstrike required the use of Google's servers or programming help from their engineers, then just-war theory holds that Google's facilities may be legitimately bombed and its personnel attacked.

3.1.3 Proportionality

Proportionality in just-war theory is the idea is that it would be wrong to use more force than necessary to achieve one's legitimate military objective, including as a punitive or deterrent response to an attack. For example, a cyberattack that causes little harm should not be answered by a conventional attack that kills hundreds (Walzer 2006; Coady 2004); that would seem to be a disproportionate response, in that less force could have achieved the same goals. This is not to say that a kinetic attack cannot be a just response to a cyberattack, depending on the severity of either. As one US official described the nation's cyberstrategy, "If you shut down our power grid, maybe we will put a missile down one of your smokestacks" (Gorman and Barnes 2011).

A challenge to proportionality is that certain cyberattacks, like viruses, might spiral out of control regardless of the attackers' intentions. While those consequences could be tolerated to prevent even worse consequences, lack of control means an attack might not be able to be called off after the victim surrenders, violating another key law of war. Such attacks thus raise issues of unintended proliferation and the possibility of widespread conflict, as attacks and counterattacks may spread beyond intended victims, undermining principles of both discrimination and proportionality. Another issue is that the target of a cyberattack may have difficulty in assessing how much damage they have received. A single malfunction in software can cause widely varied symptoms; thus a victim may think they have been damaged more than they really have, and counterattack disproportionately. Therefore, counterattack—a key deterrent to unprovoked attacks—is now fraught with ethical dilemmas.

3.1.4 Attribution

Discrimination in just-war theory also requires that combatants be identifiable to clarify legitimate targets—the principle of attribution of attackers and defenders. Terrorism ignores this requirement and therefore elicits moral condemnation. A problem with cyberwarfare is that it is very easy to mask the identities of combatants (Dipert 2010). Then counterattack risks hurting innocent victims. For example, the lack of attribution of Stuxnet raises ethical concerns because it removed the ability of Iran to counterattack, encouraging them towards ever more extreme behavior.

Attribution is an issue not only of moral responsibility but also of criminal and civil liability: we need to know whom to blame and, conversely, who can be absolved of blame. To make attribution work, we need international agreements. We first could agree that cyberattacks should carry a digital signature. Signatures are easy to compute, and their presence can itself be concealed with the techniques of steganography, so there are no particular technical obstacles to using them. Countries could also agree to use networking protocols, such as IPv6, that make attribution easier, and they could cooperate better on international network monitoring to trace sources of attacks. Economic incentives can make such agreements desirable.

3.1.5 Treacherous Deceit

Perfidy, or deception that abuses the necessary trust for the fair conduct of warfare, is prohibited by both Hague and Geneva Conventions. For instance, soldiers are not permitted to impersonate Red Cross workers and adversary soldiers. However, some ruses, misinformation, false operations, camouflage, and ambush of combatants are permissible. Cyberattacks almost inevitably involve an element of deception to make operations of a computer or network appear to be normal when they are not, as with tricking a user to click on a malicious link.

So, to what extent might cyberattacks count as perfidy and therefore be illegal given international humanitarian law (Rowe 2009)? Consider, for instance, an email virus that purports to come from the International Committee of the Red Cross: this would seem to be a reasonable analogue to the prohibited act of posing as a humanitarian worker. Similarly, an email virus that purports to come from one's own military organization would breach the same shared trust as impersonating an enemy soldier does.

The moral impermissibility of perfidy is tied to the concept of treachery, and a prototypical example of a treacherous (and illegal) act in war is to kill with poison. Yet there are poisons that can kill quickly and painlessly, much more humanly than a bullet to the head. And spraying poisons in open battle is prohibited chemical or biological warfare. This apparent paradox suggests that the concept of treachery (and therefore perfidy) is fuzzy and hard to apply. We don't get as angry when software betrays us as when people betray us. But maybe we should—software would be better if users were less complacent.

3.1.6 *What Now?*

The above issues do not exhaust the moral and philosophical controversies surrounding cyberwarfare. For instance, just-war theory also requires that wars are publicly declared by the proper authority. Yet the ambiguity of the attacker's identity is a major part of cyberwarfare's allure, as is therefore waging a secret war. These issues suggest that either we need to quickly figure out how cyberwarfare fits into the extant framework of IHL and just-war theory (Cook 2010), or if emerging capabilities to cyberattack require rewriting the rules of war (Dipert 2010). Some scholars have cast doubt that cyberwarfare is much different from previous forms of warfare or that it requires a "new ethics" (Crisp 2012; Rid 2012). Whether or not they are right, it should be clear that cyberwarfare is burdened with legal and moral hazards, some of which we described above. These hazards are perhaps solvable, but they are not solved now. And this makes "active defense" or counter-cyberattacks, at least by the nation-state, morally problematic.

3.2 Stand Your (Cyber)Ground: An Interim Solution?

If we could conduct cyberdefense outside the military frame, then we can avoid at least the legal issues above, if not also the moral ones. But do we have a good reason—other than to sidestep these issues—to use a different frame? In this section, we will suggest that we do (Lin 2012).

First, it is still an open question of whether or not military organizations should take the lead in national cyberdefense, even against foreign-based attacks. Currently in the US, a major controversy with cybersecurity legislation is whether the US Department of Homeland Security (DHS) or the US Department of Defense (DoD) should bear responsibility for defending the nation's digital borders (Jensen 2011; Jackson 2012). Reasonable arguments can be made to support and criticize either one as the lead agency for cybersecurity, but these arguments are not relevant to our discussion here. Rather, the point is simply that we are using a law-enforcement—as distinct from military—frame if we believe that DHS should take the lead, which is not unreasonable.

However, we are not proposing a law-enforcement frame here. Instead, we want to offer a third option that government need not be involved—a solution that would avoid the DHS versus DoD debate, as well as the aforementioned difficult issues related to IHL and just-war theory. This option models the "Stand Your Ground" laws in the US that are rooted in the basic human right to self-defense, and it authorizes counter-cyberattacks by private companies, which have been the main victims of harmful cyberactivities by foreign actors to date. (We will present details of this option shortly).

One reason why government need not be involved is that government is, in fact, currently not involved much at all (Riley and Walcott 2011). That is, the US government has hardly protested, much less prosecuted, the perpetrators of major cyberattacks, again with industry companies as the principal victim of such attacks. Thus, it is workable to avoid governmental intervention to the extent that the status quo of nonintervention is workable or is expected to continue anyway. To be sure, part of the reason for this inaction is the difficulty of identifying the attacker with reasonable certainty for a serious state response, such as trade sanctions or a military strike against a foreign aggressor. Nevertheless, there is little, if any, state protection for industry targets in the cyber domain.

So despite existing laws against cybercrimes and related activities, there is little enforcement of these laws, and therefore the cyber domain appears to be lawless. In that regard, a natural analogy to which we might look for consistent policy is the "Wild West" of American history. Both the Wild West and cyberspace now are marked by general lawlessness; bad guys often operate with impunity against private individuals and companies, as well as what government exists in those realms, such as the lone sheriff. The distinctively American solution to the Wild West was found in the second amendment to the US Constitution: the right to bear arms. As more private citizens and organizations carried firearms and could defend themselves, the more outlaws were deterred, and society as well as the rule of law could then stabilize and flourish.

We also find this thinking in current "Stand Your Ground" laws in the US that authorize the use of force by individual citizens. If such laws make sense, could this model work for cyberspace?

3.2.1 Why it is Reasonable

Not to endorse this solution (or "Stand Your Ground" laws) but merely to offer it for consideration as a new option: what if we authorized *commercial companies* to fight cyberfire with cyberfire? Some have already started to explore the legality of active defense (White Wolf Security 2007; Owens et al. 2009; Willson 2012), i.e., offensive operations, but let us further consider its ethical foundation here. As in the Wild West, civilians are the main victims of pernicious cyberactivities. Some estimate that industrial cyberespionage costs US companies billions of dollars a year in lost intellectual property and other harms (Goldman 2011); UK companies also report annual losses in the billions (Blitz 2012). As in the Wild West, they now look to government for protection, but government is struggling badly in this role, for the above-mentioned reasons and others. If we consider the US (or any other nation) as one member of the world community, there is no clear authority governing international relationships, and this make our situation look like a "state of nature" where no obvious legal norms exist, at least with respect to cyber.

This option isn't completely outlandish, because precedents or similar models exist for the physical, nondigital world today. In the open sea, commercial ships are permitted to shoot and kill would-be pirates (United Nations 1982). Security guards for banks are allowed to shoot fleeing robbers (e.g., New York Penal Law 2012). Again, "Stand Your Ground" laws—which give some authority and immunity to citizens who are being threatened or attacked—also operate on the same basic principle of self-defense, especially where few other options exist.

A key virtue of "Stand Your *Cyber*ground" is that it avoids the unsolved and paralyzing question of what a state's response can be, legally and ethically, against foreign-based attacks; the state is no longer involved. If the state were to make a wrong move, it could become a war crime or provide an adversary with just cause to respond with force. Again, the point of reframing cybersecurity as a nonmilitary issue isn't so much to avoid stringent but sensible requirements in IHL and just-war theory, though that might be a benefit if there are independent reasons to support a different frame.

As useful a model as it is for thinking through at least some issues in cyberwarfare, just-war theory is also limited, for at least the following reasons. First, just-war theory is most powerful when applied to state actors, particularly ones engaged in traditional warfare. This is not to say that just-war theory tells us nothing about other forms of conflict; for example, it clearly inveighs against some form of asymmetric warfare, e.g., terrorism. Rather, this gives rise to a second consideration, already discussed above: much of cyberconflict takes place among private citizens or, in many cases, between private citizens and corporations. Just-war theory would typically not be applied to this sort of dynamic, but rather we would turn to the strictures of criminal law; and, as we will see in the following, there is some promise in this regard.

3.2.2 *Controversy in "Stand Your Ground" Laws*

Of course, "Stand Your Ground" laws are not without significant controversies. In recent history, this is represented by the following criminal case in the US: On February 26, 2012, outrage broke out when George Zimmerman, a neighborhood watch coordinator, shot and killed Trayvon Martin, an unarmed 17-year old. The shooting took place in a Sanford, Florida gated community. The media fallout primarily seized upon Martin's race—he was African-American—and the failure of the police to arrest Zimmerman for several weeks. However, a secondary emphasis was on a Florida law that weighed in Zimmerman's favor, namely Florida's "Stand Your Ground" statutes (Florida Statutes 2011). Florida, though, is hardly alone in having "Stand Your Ground" provisions; many states have them, and others are currently considering them.

In understanding "Stand Your Ground", it is perhaps easiest to start with its contrary: a duty to retreat. Under the common law, self-defense is widely recognized, which is to say that one person can justifiably use (at least some) force against another if the former is in reasonable apprehension of imminent bodily injury. Unpacking this claim takes us too far afield, but let us at least briefly consider some basic features. First, the person invoking self-defense need not be in *actual* apprehension of imminent bodily injury; so long as the apprehension is *reasonable*, it is sufficient to mitigate liability, either criminal or civil. Second, the apprehension needs to be imminent, which is to say that self-defense cannot be used against threats or provocations. Third, the force used cannot be excessive; rather, it can only be what would be reasonable to prevent injury.

Intuitively, self-defense protections strike most of us as eminently plausible: we hardly expect people to suffer preventable injuries at the hands of others. The key to understanding "Stand Your Ground", though, is in recognizing that it provides even greater protections to those who wield protective force than does the traditional doctrine of self-defense. Specifically, this distinction trades on the duty to retreat. Under self-defense, if the person being attacked could have escaped without injuring his assailant, he is usually expected to do so; if he does not, he may be found liable for the injuries that he causes. "Stand Your Ground", however, is more forgiving insofar it does not require the attacker to exercise the option of retreat before using force.

Surely we can understand why reasonable retreat would be required, so why does "Stand Your Ground" jettison it? Progenitors to contemporary statutes ran under the "Castle doctrine", which provides extra protections for a person's residence and has been widely adopted. Under this doctrine, whether a person has a duty to retreat depends on where he is, and the duty is absent when he is in his own house. In fact, the "Stand Your Ground" locution originated in a case deriving from just this sort of situation: "[the homeowner] may stand his ground, and, if need be, kill his adversary" (Beard v. US 1895). In the contemporary legislative landscape, "Stand Your Ground" extends beyond just domestic contexts. The basic rationale for this expansion is one that Oliver Wendell Holmes expressed a long ago, "detached reflection cannot be demanded in the presence of an uplifted knife" (Brown v. US 1921); in other words, the duty to retreat is simply unfair to the person who is attacked.

3.2.3 How it Could Work

Our exploration—but not necessarily an endorsement—of a "Stand Your *Cyberground*" policy starts from a similar assumption of a basic right to self-defense, as found in extant "Stand Your Ground" and other laws. And as imperfect as any analogy inevitably is, but nonetheless useful (Hollis 2008), there are important similarities here. In both cases, the victim does not have access to government protection, for all practical purposes: in the home-invasion case where seconds matter, that the police may be minutes away is little consolation or protection; and in the corporate cyberattack case where there is no prosecution, that we have laws against cyberattacks are also of little help. In both cases, there's nowhere to reasonably retreat, even if there were such a duty to retreat. Even considering some of the other analogies proposed for cyberspace—e.g., outer space and Antarctica—it's reasonable to assume that something like "Stand Your Ground" would also apply in those lawless frontiers, if an attack were to occur against a private party there. This also suggests a correlative policy that at least some cyberattacks, perhaps even between nation-states, should be treated as "frontier incidents" rather than the more serious "acts of war", to the extent that cyberspace is still a frontier (Watts 2011; Schmitt 2010).

Where "frontier justice" may evoke images of brutal eye-for-an-eye retaliation, or *lex talionis*, this need not be the case for cyberpolicy. A counter-cyberstrike by a defending company does not have to be as dramatic as the initial attack or anything else we usually associate with an "attack." For instance, the response could be to forcibly install software patches and anti-malware applications on an attacking "botnet" or network of zombie computer systems, usually hijacked without their owners' knowledge; or it could be to encrypt an attacking computer's data and operating system until some remedy is achieved; or, as Microsoft had done in 2012, the response could be to render a botnet inoperable (Infosec Island 2012). Other remedies include creating a "honeypot" or diversionary target (Rowe et al. 2007), e.g., a fake directory of trade secrets, in order to misdirect the cyberattacker, plant false information for attackers to "discover", keep attackers occupied to buy time for defense and evidence-collection, and other ends. Compare these to decoys, mock operations, camouflage, and other tactics that militaries and intelligence agencies are permitted to conduct to mislead adversaries.

If we like, "Stand Your Cyberground" could require a judicial warrant prior to a cyber-counterstrike, that is, *ex ante* justification or authorization before the event. However, this may be unnecessary, since there could be also *ex post* justification, that is, authorization in virtue of an initial attack. Again, in open-seas piracy and other scenarios today, a victim does not need to request approval prior to defending itself with a counterattack. As further safeguards, the state (or industry, to avoid the state's involvement) may require that counter-cyberattacks be reported, either before or after the fact, to ensure there is reasonable cause in those actions, or else face some penalty for negligence or other deficiencies.

As with the counterattacking of pirates, a cyber-counterattack has many potential benefits, including neutralizing the threat, deterring future threats, and providing some measure of justice, in contrast to doing nothing. Further, where initial cyberattacks are often anonymous or conducted through an unwitting proxy, a counterstrike on an "innocent" third-party's system—say, computers owned by China but hijacked and used for an attack by unknown hackers—could elicit pressure from the third-party (in this case, China, a nation of significant influence) to identify the real aggressor. (We will say more about the innocence of these third-parties below.) Short of a sound or responsible national cyberpolicy that accounts for IHL and just-war theory, a counterstrike outside the military frame helps to avoid a larger cyberwar as well as kinetic war. If this is still unsatisfying or unsustainable, then "Stand Your Cyberground" may help motivate lawmakers to more quickly develop a sensible national cyberpolicy.

3.3 Possible Objections and Replies

Here we briefly consider several objections to the "Stand Your Cyberground" policy, as it is undoubtedly controversial. In the process, we clarify how such a proposal might work, in case it is ultimately defensible. Again, this is not an exhaustive list of objections but only some immediate worries, which may be overcome to make a *prima facie* case for "Stand Your Cyberground."

3.3.1 Only the State has a Monopoly on Violence

Objection: Only the state can engage in war or otherwise violent actions; companies legally cannot, as governments have a legitimate monopoly on warfare and violence.

Reply: There are certainly areas in which government intervention is required to regulate or even supplant private interactions; political parties routinely argue over the appropriate extent of such government usurpation of individual sovereignty. But almost everyone agrees that government should have the sole legitimate use of violent force against other people. The most basic argument for this requirement is that vigilante justice runs into a regression problem, when friends or loved ones of private individuals retaliate for their loved one's murder, and then the loved ones of the original transgressor return the favor, and on and on as some legendary family and ethnic feuds have continued.

If governments must have a monopoly on the legitimate use of violence, must they also have the sole legitimate use of cyberattacks? No, not necessarily. To say that a state has a monopoly on violence seems to imply it is capable of inflicting violence or otherwise enforcing laws so that individuals need not resort to violence themselves. With industry cyberattacks, if the state does have this power, it has not been exercising it, as justice may demand. Again, part of the problem is that it's difficult to identify the aggressor, as ethics generally would seem to require; so this is not so much the state's fault as it is the nature of cyberattacks. Nevertheless, the state is not living up to its implicit promise to protect its citizens, which was the basis for claiming a monopoly on violence. Further, it is not true that governments claim a monopoly on violence, to the extent that they allow commercial ships to defend themselves against pirates, or bank security guards to shoot fleeing robbers, or private citizens to counterattack given "Stand Your Ground" laws.

If the objection, however, is that only the state has the power to wage war, then this begs the question at hand: we have suggested that a cyberconflict does not need to be viewed through the lens of war. Suppose a cross-border kinetic attack occurs on a bank (or your house): the bank (or you) would seem to have a reasonable claim to defend itself from such attacks, including with deterrent force, especially if government is unresponsive. This is not a war-powers problem but one of basic self-defense.

3.3.2 Only the State has the Resources to Counterattack

Objection: Related to the above, many companies are typically not big enough to mount an effective counterattack. As a matter of simple utility and following the principle of division of labor, even if companies could handle cyber-counterattacks, government still should handle all cyberattacks, given its considerable resources and economies of scale.

Reply: Companies need not act alone; they could form consortiums or cooperatives to gather resources and expertise for cyber-counterattacks, if the individual company lacks resources. Or they could simply outsource the job to a third-party with cyberdefense as its core competency or product, as a bank might hire private security services. Such voluntary solutions appear to be better than involving governments, insofar as state-sponsored attacks increase the risk of formal war. Further, decentralizing this function distributes our own targets for attacks, e.g., rather than having a central government agency as a single target, an adversary could have to contend with many private organizations, if it wants to knock out cyberattack capabilities. Decentralizing this function also allows for greater diversity of solutions, with nationally and internationally recognized "best practices" emerging over time. A robust corporate culture for problem-solving can be generally preferable to government intervention, especially when that intervention could mean kinetic (and not merely cyber) war.

3.3.3 There's Still the Problem of Attribution

Objection: There is a great risk of misattribution in cyber-counterattacks, potentially with innocent third parties being harmed. Even if IHL is not violated by industry-sponsored counter cyberattacks, it is still immoral to attack a party without first identifying it and ensuring that it is the actual aggressor. For instance, botnets are a common form of attack, but they're victims too, not the real aggressor.

Reply: Attribution may be a red herring here. For example, the US knows China has repeatedly cyberattacked it, but the US doesn't want to do anything about it, because there are bigger political and economic issues it wants to negotiate. Even if the US doesn't "know" this, it seems to have good reason to think so (Riley and Walcott 2011). Further, there is a widespread consensus that clear attribution is not required when sailors defend against pirates, or homeowners against robbers, and so on. It is enough to know that one is being attacked and is defending oneself against the attack, even if the attacker is not the actual aggressor, e.g., if the pirate or bank-robber was really a coerced father whose family was taken hostage and threatened to be killed by true bandits.

As for innocent third parties and botnets (innocent computers hijacked by others to commit cybercrimes): again, even if we know that a pirate was really an innocent fisherman whose family was being held hostage, the fact remains that the pirate poses a threat to the safety of the targeted ship and its crew and passengers. It is therefore still not unreasonable to neutralize the threat by counterattacking the pirate, even if we know there is a puppet-master elsewhere who is responsible for the pirate's actions. Similarly, it would seem reasonable to counter-cyberattack a third party who we believe was coerced or otherwise not complicit in their initial attack.

Where we may choose to use less-than-lethal means against a fisherman we know to be an unwilling pirate, we may likewise choose less dramatic means in a counter-cyberattack. Again, such a counterattack need not be crippling or highly damaging, e.g., if it merely forces an anti-malware installation. If the cyberdefense routinely inoculates and removes malware from consumer machines, such an "attack" could actually be a great benefit to the wired world, as well as a more effective general solution to cyberattacks. This is to suggest that we may understand botnets with the public-health model of bioethics: In cases of infectious diseases, such as typhoid, patient autonomy is secondary to stopping the disease that threatens many others (Leavitt 1997). Likewise, botnets are a public-health hazard too in a sense; and even if the owners of botnet computers are not complicit in the attack and want to refuse an inoculation, the overriding greater good of public health can reasonably trump that innocent autonomy.

Botnets, however, are less innocent than the unwilling pirate above in an important sense. One can argue that the hijacked computers comprising a botnet still bear some responsibility for cyberattacks (Owens et al. 2009, p. 210). For instance, responsible owners of those computers could be said to have some positive obligation to install antivirus software and otherwise exercise due diligence in ensuring responsible use of their machines; failing to do so puts the computers at risk of becoming hijacked and used for pernicious ends. In the bioethics model, this is analogous to something like careless or oblivious patients who don't take reasonable precautions as they enter a zone of infection; this lack of reasonable diligence weighs against their right to autonomy.

3.3.4 Counterattacks will Escalate Conflict

Objection: Cyber-counterattacks will only encourage the escalation of conflict. Violence begets more violence, so we should forgo a counterstrike option in favor of some other response.

Reply: Perhaps, though this is a general objection to any response to aggression, whether a kinetic war, cyberconflict, or a schoolyard fight. Any response—even a nonresponse—may encourage the aggressor on. This seems true for cyberconflicts, even with a national cyberpolicy in place. Note that diplomacy and negotiations may be impossible in cyberconflicts, if the victim does not know the identity of the attacker, i.e., with whom one ought to negotiate.

Insofar as deterrents work, what seems to be clear is that a nonresponse is not a deterrent. A "Stand Your Cyberground" solution could be an immediate deterrent and pressure "innocent" third-parties to help find the real aggressor for compensation and/or punishment. Further, understanding how cyberattacks occur may help us to take our computing practices more seriously and generally replace the naivete common today with a more sophisticated relationship that ultimately could engender greater, not lesser, trust. It seems possible that the current asymmetry of possible harm between elite hackers and average citizens could gradually be replaced with a grudging trust built on the possibility of mutually assured harm from cyberattacks, and hence act as a long-run general deterrent to cyberattacks; when hacking involves a considerable risk of counterattack to the hacker, it's entirely possible less hacking will result.

Hence, though the worry about escalation is reasonable (no matter what policy is adopted), ultimately it becomes an empirical question. Looking at the American debate on whether we should allow more people to carry guns, one criticism is that it'd escalate violence, especially accidental and wrongful shootings; however, others predict that more guns will force us to be more civil and therefore reduce violence, since we wouldn't want to risk offending an armed person (Debatepedia 2011). This was supposedly the case in the Wild West, which we suggested was an analogy to our current situation in the cyber domain. Where "Stand Your Cyberground" differs from the debate on guns is that there'd be little danger of an industry company launching a cyberattack by accident or without cause, like a careless, emotional, or angry gun-owner might shoot someone. Designing and implementing a complicated cyberattack is not typically an impulsive gesture. But that capacity would still remain a deterrent to others: to not cyberattack a company that could plausibly respond in kind.

The failure to defend oneself also risks escalation. After all, failing to respond to a cyberattack is an incentive for hackers to continue, if not escalate, their activities. This reasoning lies behind zero-tolerance policies for minor urban crimes and helps explain why bad, crime-ridden neighborhoods tend to get worse: because the perpetrators have no incentive to discontinue their assaults, given the absence of reliable law enforcement or self-defense. It is unclear how doing nothing will de-escalate a cyberconflict: a hacker is not like the angry drunk who will eventually run out of steam and pass out or sober up. If cyberattacks are still profitable, then they will continue or increase.

3.3.5 Malicious and Ideological Hackers will not be Deterred

Objection: Even if financially motivated hackers can be deterred or expected to not exact revenge, this may not be the case with malicious or ideological hackers, such as Anonymous. Rather, a cyber-counterattack may instead play into a hacker's agenda of anarchy.

Reply: Perhaps, but this may create political will to fight cybercrimes, if the cyber domain devolves into a Wild West—a drastic but necessary catalyst for action. And as major organizations worldwide, such as Amazon.com and various credit card companies, discovered after being attacked by Anonymous, the alternative of doing nothing seems worse. Would hackers retaliate if a company were to pull out its cybergun? Maybe if they were motivated by revenge, but again, like the average mugger, the motive in the end is usually primarily financial, even if some hackers and hooligans do it for fun. Anonymous hacked in support of WikiLeaks precisely when Amazon et al. were denying donations to WikiLeaks. Even ideological hackers need funds. And so eventually even the members of groups like Anonymous can be harmed by cyber-counterattacks, especially counterattacks that impose financial or technological hardships on the original hacker.

"Stand Your Cyberground" has the virtue of advertising to would-be attackers, whatever their motivation, that industry is not an easy target, and this has deterrent value. Perhaps some hackers will take that as a challenge, but they're not so much the rational adversary (who are motivated by profit) that this policy is meant to address. Just as some hackers and muggers may strike back harder if the victim resists or fights back, this minority group shouldn't drive policy that's otherwise reasonable and potentially more helpful than not. Of course, a rational hacker could preemptively declare a policy of striking harder if a company resists, as a way to deter deterrence, but again this would seem to be an even smaller segment of that community, and we shouldn't let these outlier (and theoretical) cases drive policy for the larger world.

3.3.6 Even if IHL is not Violated, other International Laws may be

Objection: Given that many companies are multinational, their counter-cyberattacks may violate other aspects of international law, even if not in violation of IHL. Conceivably, it could open the company up to international prosecution.

Reply: There's irony in prosecuting a defending company that counterstrikes but not the initial aggressor, so it's unclear what the political appetite would be for such prosecutions. If cyberattacks come to be routinely prosecuted internationally, then there would be a major step towards leaving behind the "Wild West" of current cyberconflict and moving toward international regulations, greatly obviating the need for a "Stand Your Cyberground" policy. But that would require the prosecution be carried out in such a way that companies no longer need to actively defend themselves from cyberattack, and such a vista remains distant at best. It remains hard to envisage a thoroughgoing and extensive enough international consensus on cyberlaw that could render private companies and individuals in as little need of cyberdefense as average citizens do against shootings. When assaults are common and hard to police, one must expect people to begin to actively defend themselves.

Further, it is hard to prosecute a company without clear attribution—and in principle, companies could respond in their cyber-counterattacks as anonymously as its attackers do (perhaps, if feasible, even using the same botnets), as that strategy seems to be effective for attackers. If computer forensics advances to the point in which there is a robust system for identifying and reliably attributing cyberattacks (and settled international cyberlaw for discriminating illegitimate attacks from other cyberactivities), then our proposal will be no longer needed, and attackers can be identified and prosecuted, i.e., cyberlaw can actually be enforced.

3.3.7 A Judicial Process Implies State-Sponsorship of "Stand Your Cyberground"

Objection: Requiring a judicial warrant or reporting of cyber-counterstrikes amounts to state-sponsorship for the "Stand Your Cyberground" policy (Owens et al. 2009, p. 211). As with states that turn a blind eye toward terrorists within their own borders, states can reasonably be blamed for any cyber-counterstrikes. This means the policy does not reduce the risk of war after all.

Reply: First, it may be the case that cyber-counterattacks could proceed without any judicial oversight at all; that would be the most *laissez faire* version and would presumably obviate any risk of war from the "Stand Your Cyberground" policy. After all, other kinds of ritualized exchanges of harm, often even those involving kinetic violence, do not threaten to lead to war, e.g., gang violence across international borders. Cross-border cyberattacks and counterattacks would be more problematic, but as we suggested above, it makes no sense to prosecute a cyber-counterattack when the initial attacker goes unpunished. It remains plausible that transnational disputes will result from such counterattacks, but there is no reason to think they are more likely to lead to war than other types of international crime, particularly cybercrime, that already exist. Indeed, if the "Stand Your Cyberground" policy does become a credible deterrent and reduce international hacking, it may well defuse international tensions, not raise them.

If the government does become involved in cyber-counterattacks to the limited extent of requiring *post hoc* notification or *ex ante* warrants, things become more complicated. But the end result remains the same: there are multiple venues to appeal the legal findings of one country to a higher court, beginning with low-level government to government negotiations and culminating with appeals to the United Nations and the International Criminal Court. None of those involves war, and it is hard to imagine a cyber-*counter*attack—which assumes a cyberattack causing harm already took place—in which the counterattack by itself precipitated war. Cyber-counterattacks are unlike terrorism in that they are a specific response to a specific injury, in kind, and without larger political goals beyond self-defense. If nation-states begin a "first strike" cyberattack policy, that may well constitute war or an incitement to war, but that goes well beyond what "Stand Your Cyberground" is envisioned to achieve.

3.3.8 Industry Counterattacks may Destroy Evidence Needed for Prosecution

Objection: If we allow victims to unilaterally counter-cyberstrike, that will likely contaminate or destroy evidence needed to prosecute the initial (and presumably illegal) cyberattack (Owens et al. 2009, p. 206; Infosec Island 2012).

Reply: First of all, what prosecution? Even if prosecution of the aggressor were forthcoming, this is a problem for any act of self-defense. For instance, by allowing commercial ships to repel pirates, we risk destroying evidence on the alleged pirate's unlawful activities; by allowing individuals to counterattack assailants, we risk destroying evidence that would convict the alleged aggressors. But as real as this risk is, prosecution is secondary to self-defense and limiting the harm of the initial attack. Allowing a cyberattack to continue for the sake of a possible prosecution makes as much sense as letting a suspicious fire to keep burning so to not disturb evidence that may convict an arsonist. Further, in regulating "Stand Your Cyberground", the state or industry could require capturing and filing relevant data related to the initial attack, perhaps deploying independent emergency-response teams to document the initial attack.

3.3.9 Cyberwarfare Doesn't Raise New Issues

Objection: Do cyberattacks really raise new moral issues? They seem to be merely old ethical issues in a new technological dress (Crisp 2012).

Reply: Given "ought implies can", as new technologies emerge with new capabilities, novel ethical questions ineluctably arise. Moral dilemmas over killing and letting die and even organ transplantation arose once medical technology forced us to redefine death: the case of Terri Schiavo would not have been an issue two centuries ago (Caplan 2005). Whenever such technological developments change the concepts currently in use, they likewise inevitably challenge our received ethics. Just-war theory is challenged by the rise of semi-autonomous robots: do drone strikes mean that the US is at war with Yemen, or not? Similarly, the distinctive nature of cyberattacks, whose very nature upends the traditional notion of kinetic force as required for attack, places extreme tension on just-war theory, law enforcement, or any other traditional frame for assessing their ethics. Hence, we believe new ethical—and philosophical (Taddeo 2012)—issues are raised by cyberattacks, and so until and unless policymakers come to grips with regulating this novel form of aggression, it falls on private individuals to work out a *modus operandi* for this new reality.

3.4 Conclusion

How we justify and prosecute a war matters. For instance, the last US presidency proposed a doctrine of preventive or preemptive war, or the "Bush doctrine": if a nation knows it will be attacked, why wait for the damage to be done before it retaliates (Tierney 2011)? But this policy breaks from the just-war tradition, which again historically gives moral permission for a nation to enter war only in self-defense. With the Bush doctrine, the US seeks to expand the triggers for war, but this could backfire spectacularly. For instance, Iran reports contemplating a preemptive attack on the U.S. and Israel, because it believes that one or both will attack Iran first (BBC 2012). Because intentions between nations are easy to misread, especially between radically different cultures and during political elections, it could very well be that the US and Israel are merely posturing as a gambit to pressure Iran to open its nuclear program to international inspection. However, if Iran were to attack first—with either kinetic or cyber means—it would seem hypocritical for the US to complain, since the US already endorsed the same policy of first strike (Wright 2012).

A key problem with a first-strike policy is that there are few scenarios in which we can confidently and accurately say that an attack is imminent. Many threats or bluffs that were never intended to escalate into armed conflict can be mistaken as "imminent" attacks. This epistemic gap in the Bush doctrine introduces a potentially catastrophic risk: that nation delivering a preemptive or preventative first strike may turn out to be the unjustified aggressor and not the would-be victim, if the adversary really was not going to attack first. Further, by not saving war as a last resort—after all negotiations have failed and after an actual attack, a clear act of war—the Bush doctrine opens the possibility that the US (and any other nation that adopts such a policy) may become ensnared in avoidable wars. At the least, this would cause harm that otherwise might not have occurred to the warring parties, and it may set up an overly stretched military for failure, if battles are not chosen more wisely.

Here's the relevance to cyberwarfare: Our world is increasingly wired, with new online channels for communication and services interwoven into our lives virtually every day. This also means new channels for warfare. Indeed, a target in cyberspace is more appealing than conventional physical targets, since the aggressor would not need to incur the expense and risk of transporting equipment and deploying troops across borders into enemy territory, not to mention the political risk of casualties. Cyberweapons could be used to attack anonymously at a distance while still causing much mayhem, on targets ranging from banks to media to military organizations. Thus, cyberweapons would seem to be an excellent choice for an unprovoked surprise strike.

Today, many nations have the capability to strike in cyberspace—but should they? The laws of war, or IHL, were not written with cyberspace in mind. So we face a large policy gap, which organizations and experts internationally have tried to address in recent years (e.g., Owens et al. 2009; Lieberthal and Singer 2012; Libicki 2009; H. Lin 2012). But there is also a gap in developing the ethics behind policies, as we described in the first section above. As an interim solution, we suggest a reframing of the cybersecurity discussion away from the military frame, i.e., away from the nation-state level, and more toward the private-defense frame, i.e., closer to the individual-actor level.

This reframing seems defensible, given related legal precedents. And, separately, it offers many benefits, including some measure of justice to victims, deterrence for aggressors, and so on. While we offer this "Stand Your Cyberground" policy as a prelude to a more complete discussion of its feasibility, we should also note that it is already being adopted by companies right now: "Frustrated by their inability to stop sophisticated hacking attacks or use the law to punish their assailants, an increasing number of US companies are taking retaliatory action" (Menn 2012; Infosec Island 2012). So regardless of whether the policy is prudent or ethical, it is apparently already a *de facto* policy for some, and this makes an examination of its details—including how it could responsibly proceed—all the more urgent.

References

Allhoff, F. 2012. *Terrorism, ticking time-bombs, and torture*. Chicago: University of Chicago Press.

Arquilla, J. 2012. Cyberwar is already upon us. Foreign policy (March/April). http://www.foreignpolicy.com/articles/2012/02/27/cyberwar_is_already_upon_us?page=full. Accessed 22 October 2013.

Aquinas, T. 1948. *Summa Theologica* (Trans. Fathers of the English Dominican Province). New York: Benziger Books.

BBC. 2012. Iran Says Preemptive Strike on 'Enemies' Possible. BBC News, February 21. http://www.bbc.co.uk/news/world-middle-east-17116588. Accessed 22 October 2013.

Beard v. United States. 1895. 158 U.S. 550, 563 (1895). http://caselaw.lp.findlaw.com/scripts/getcase.pl?court=us&vol=158&invol=550. Accessed 22 October 2013.

Blitz, J. 2012. MI5 chief sets out price of cyber attack. Financial times. http://www.ft.com/cms/s/0/a970810c-bef2-11e1-8ccd-00144feabdc0.html#axzz1zDB9cncm. Accessed 22 October 2013.

Brown v. United States. 1921. 256 U.S. 335, 343. (1921). http://supreme.justia.com/cases/federal/us/256/335/case.html. Accessed 22 October 2013.
Caplan, A. 2005. The time has come to let terri schiavo die. MSNBC.com, March 18. http://www.msnbc.msn.com/id/7231440/ns/health-health_care/t/time-has-come-let-terri-schiavo-die/#.T-4VArWe4aw. Accessed 22 October 2013.
Clarke, R. 2010. *Cyber war: The next threat to national security and what to do about it*. New York: Ecco.
Coady, C. A. J. 2004. Terrorism, morality, and supreme emergency. *Ethics* 114: 772–789.
Cook, J. 2010. 'Cyberation' and just war doctrine: A response to randall dipert. *Journal of Military Ethics* 9 (4): 411–423.
Crisp, R. 2012. Cyberwarfare: No new ethics needed. Practical ethics: Ethics in the news. http://blog.practicalethics.ox.ac.uk/2012/06/cyberwarfare-no-new-ethics-needed/. Accessed 22 October 2013.
Debatepedia. 2011. Debate: Gun control debatepedia index. http://dbp.idebate.org/en/index.php/Debate:_Gun_control. Accessed 22 October 2013.
Dipert, R. 2006. Preventive war and the epistemological dimension of the morality of war. *Journal of Military Ethics* 5 (1): 32–54.
Dipert, R. 2010. The ethics of cyberwarfare. *Journal of Military Ethics* 9 (4): 384–410.
Florida Statutes. 2011. Chapter 776: Justifiable use of force. The Florida Senate. http://www.flsenate.gov/Laws/Statutes/2010/Chapter776/All. Accessed 22 October 2013.
Goldman, D. 2011. The cost of cybercrime. CNNMoney, July 22. http://money.cnn.com/galleries/2011/technology/1107/gallery.cyber_security_costs/index.html. Accessed 22 October 2013.
Gorman, S. and J. E. Barnes. 2011. Cyber combat: Act of war. *The Wall Street Journal: Technology*, May 30. http://online.wsj.com/article/SB10001424052702304563104576355623135782718.html?mod=googlenews_wsj. Accessed 22 October 2013.
Hollis, D. 2008. New tools, new rules: International law and information operations. In *The message of war: Information, influence and perception in armed conflict*, eds. G. David and T. McKeldin. Temple University Legal studies research paper no. 2007–15. http://ssrn.com/abstract=1009224. Accessed 22 October 2013.
Infosec Island. 2012. Microsoft dismisses zeus botnet takedown criticism. Infosecisland. http://www.infosecisland.com/blogview/21036-Microsoft-Dismisses-Zeus-Botnet-Takedown-Criticism.html. Accessed 22 October 2013.
Jackson, W. 2012. DOD vs. DHS: Who should mind the US' Cyber Defense? *Defense Systems*, March 27. http://defensesystems.com/articles/2012/03/27/cyber-defense-hearing-mccain-slams-dhs-favors-dod.aspx. Accessed 22 October 2013.
Jensen, E. 2011. President Obama and the changing cyber paradigm. William Mitchell law review, vol. 37, no. 5049. http://ssrn.com/abstract=1740904. Accessed 22 October 2013.
Leavitt, J. W. 1997. *Typhoid mary: Captive to the public's health*. Boston: Beacon Press.
Lieberthal, K., and P. W. Singer. 2012. Cybersecurity and US-China Relations. The brookings institution, 21st century defense initiative. http://www.brookings.edu/~/media/research/files/papers/2012/2/23%20cybersecurity%20china%20us%20singer%20lieberthal/0223_cybersecurity_china_us_lieberthal_singer_pdf_english. Accessed 22 October 2013.
Libicki, M. 2009. Cyberdeterrence and cyberwar. The RAND corporation. http://www.rand.org/pubs/monographs/MG877.html. Accessed 22 October 2013.
Lin, H. 2012. Arms control in cyberspace: Challenges and opportunities. *World Politics Review*, March 2012.
Lin, P. 2012. 'Stand your cyberground' law: A novel proposal for digital security. *The Atlantic*, April 30. http://www.theatlantic.com/technology/archive/2012/04/stand-your-cyberground-law-a-novel-proposal-for-digital-security/256532/. Accessed 22 October 2013.
Lin, P., F. Allhoff, and N. C. Rowe. 2012. War 2.0: Cyberweapons and ethics. *Communications of the ACM* 55 (3): 24–26.
Markoff, J. 2008. Before the Gunfire, Cyberattacks. *The New York Times: Technology*, August 12. http://www.nytimes.com/2008/08/13/technology/13cyber.html. Accessed 22 October 2013.

McMahan, J. 2009. *Killing in war*. Oxford: Oxford University Press.
Menn, J. 2012. Hacked firms fight back with vigilante justice. *The Globe and Mail*, June 18. http://www.theglobeandmail.com/technology/tech-news/hacked-firms-fight-back-with-vigilante-justice/article4321501/. Accessed 22 October 2013.
New York Penal Law. 2012. Article 35: Defense of justification. New York Laws. http://ypdcrime.com/penal.law/article35.htm. Accessed 22 October 2013.
Orend, B. 2005. War. The stanford encyclopedia of philosophy. Stanford University. http://plato.stanford.edu/entries/war/. Accessed 22 October 2013.
Owens, W., K. Dam, and H. Lin, eds. 2009. Technology, policy, law, and ethics regarding U.S. Acquisition and use of cyberattack capabilities. The National Academies Press. http://www.nap.edu/catalog.php?record_id=12651#orgs. Accessed 22 October 2013.
Reichberg, G. M., H. Syse, and E. Begby, eds. 2006. *The ethics of war: Classic and contemporary readings*. Malden: Blackwell Publishing.
Rid, T. 2012. Think again: Cyberwar. Foreign policy. http://www.foreignpolicy.com/articles/2012/02/27/cyberwar?page=full. Accessed 22 October 2013.
Riley, M., and J. Walcott. 2011. China-based hacking of 760 companies shows cyber cold war. Bloomberg, December 14. http://www.bloomberg.com/news/2011-12-13/china-based-hacking-of-760-companies-reflects-undeclared-global-cyber-war.html. Accessed 22 October 2013.
Rowe, N. 2009. The ethics of cyberweapons in warfare. *International Journal of Cyberethics*. 1:20–31.
Rowe, N., E. J. Custy, and B. T. Duong. 2007. Defending cyberspace with fake honeypots. *Journal of Computers* 2:25–36.
Sanger, D. 2012. Obama order sped up wave of cyberattacks on Iran. New York Times, June 1. http://www.nytimes.com/2012/06/01/world/middleeast/obama-ordered-wave-of-cyberattacks-against-iran.html?pagewanted=all. Accessed 22 October 2013.
Schmitt, M. 2010. Cyber operations in international law: The use of force, collective security, self-defense, and armed conflicts. The National Academies Press. http://www.nap.edu/openbook.php?record_id=12997&page=151. Accessed 22 October 2013.
Schneier, B. 2010. The story behind the stuxnet virus. Forbes, October 7. http://www.forbes.com/2010/10/06/iran-nuclear-computer-technology-security-stuxnet-worm.html. Accessed 22 October 2013.
Taddeo, M. 2012. Information warfare: A philosophical perspective. *Philosophy and Technology* 25:105–120.
Tierney, D. 2011. We don't need an Obama doctrine. The Atlantic, October 23. http://www.theatlantic.com/international/archive/2011/10/we-dont-need-an-obama-doctrine/247207/#. Accessed 22 October 2013.
United Nations. 1982. United nations convention on the law of the seas, especially Part VII. http://www.un.org/Depts/los/convention_agreements/texts/unclos/closindx.htm. Accessed 22 October 2013.
Walzer, M. 2006. *Just and unjust wars*. New York: Basic Books.
Watts, S. 2011. Low-intensity computer network attack and self-defense. *International Law Studies*. 87:59–87. http://www.law.berkeley.edu/files/watts–low_intensity_computer_network_attack.pdf. Accessed 22 October 2013.
White Wolf Security. 2007. Offensive operations in cyberspace. White Wolf Security Publications. http://www.whitewolfsecurity.com/publications/offensive_ops.php. Accessed 22 October 2013.
Willson, D. L. 2010. When does electronic espionage or a cyber attack become an 'Act of War'?. *Information Systems Security Association (ISSA) Journal* 8:20–24.
Willson, D. L. 2012. Hacking back in self-defense: Is it legal? Should it be?. *Information Systems Security Association (ISSA) Journal*. 10:7–10.
Wright, R. 2012. President Obama's hypocrisy on cyberattacks. The Atlantic, June 3. http://www.theatlantic.com/international/archive/2012/06/president-obamas-hypocrisy-on-cyberattacks/258016/. Accessed 22 October 2013.

Chapter 4
Technology, Information, and Modern Warfare: Challenges and Prospects in the 21st Century

Wayne McCormack and Deen Chatterjee

Abstract In international law there are two long-recognized conflicts: one between self-determination and non-intervention, and the other between self-defense and non-intervention. In Sect. I, Wayne McCormack examines the first conflict in the context of informational warfare, concluding that supplying information (or misinformation) in a foreign conflict with the objective of altering the course of the conflict is within the acknowledged sovereignty rights of a state and does not violate the non-interference right of the state in conflict. In Sect. II, Deen Chatterjee examines the other conflict—that between self-defense and non-intervention. He claims that the provision of preventive war in self-defense can get unduly interventionist, especially in the context of cyber warfare, making the world less secure. To counter this prospect, Chatterjee suggests that countries should promote prevention in non-interventionist terms by relying on the soft power of diplomacy and collaboration.

The traditional debates of war and peace have become a major focus of controversy in response to the changing nature of warfare in the twenty-first century, putting in sharp focus the issues of traditional paradigms and their limits, the moral hazards of military response, and the future of warfare. All these have vast implications for international law, justice, and human rights. This chapter looks at one important aspect of the changing terrain of today's war: the normative and legal challenges of information and communication technologies in modern warfare.

The chapter is divided into two sections. In Sect. I, Wayne McCormack examines the moral and the legal implications of informational warfare related to interfering in the internal affairs of a nation facing armed uprising or going through similar violent turmoil. McCormack's focus is primarily the turmoil of the Middle East. He discusses the benign use of information in conflict zones to alter the outcome. In Sect. II, Deen Chatterjee examines the moral challenges of the growing reliance

W. McCormack (✉) · D. Chatterjee
University of Utah, Salt Lake City, USA
e-mail: wayne.mccormack@law.utah.edu

D. Chatterjee
e-mail: deen.chatterjee@law.utah.edu

L. Floridi, M. Taddeo (eds.), *The Ethics of Information Warfare*,
Law, Governance and Technology Series 14, DOI 10.1007/978-3-319-04135-3_4,

on information and communication technologies in modern warfare. Specifically, he looks at the specter of "virtual warfare" in the blurring of the distinction between preemption and prevention in wars of self-defense.

4.1 Introduction

The phrase "informational warfare" covers a host of possibilities—attacks on another entity's information (cyber warfare), promotion of reasonably accurate propaganda into another country (Radio Free America), promotion of disinformation (telling populace of impending disasters), and financial support of candidates in elections. For example, if we send misinformation about the Assad regime into the public domain in Syria, that's just propaganda or what was once called "psy ops". At the other extreme, covert infiltration into the rebel groups with training materials could be illegal under the Nicaragua decision.[1] In between those extremes, funneling government money into a political campaign in another country is highly questionable (as an intervention into internal affairs of another sovereign) but is done almost certainly as a routine matter.

I assume that the U.S. poured substantial money and personnel into the Arab Spring of 2011. I also assume that we would have supported any candidate who ran against Hugo Chavez. These efforts parallel what the U.S. Supreme Court has authorized in holding that corporations have a constitutional right to pump all the money they want into political campaigns. And most observers assume that includes "foreign" corporations because how can you tell the difference between foreign and domestic in the global economy? We can also assume that the Court would hold that the First Amendment protects corporations in funneling money into foreign political actions. One might try to distinguish taking federal government money for that purpose but it is not easy to see a rational distinction between using government money as a contractor and using money derived from other revenue sources.

The question then becomes whether that interference in the affairs of other nations can somehow be justified under international law. But first let's explore a bit more about the content or tactics of "informational warfare".

In my lifetime, I have seen the War on Poverty (I don't recall that LBJ tried to justify the shooting of homeless persons), the War on Crime (some in the Nixon years might have tried to justify shooting criminals), the War on Drugs (I remember one Navy captain asking if he was supposed to shoot pharmacists), and now the War on Terror (I honestly don't know how you shoot a feeling). As a rhetorical flourish, the word "war" is useful for mobilizing resources. As a legal concept, however, it has very important and detailed consequences. Now people are starting to talk about "informational warfare" as if it were different from the propaganda campaigns of the past. I can think of some TV channels that might be worth destruction but that would probably fit into the category of MOOTW (military operations other than war) because it would not be a prolonged conflict.

[1] Nicaragua v. United States, I.C.J. Reports 1986, p. 14.

In recent years, a rather Orwellian notion has arisen around the concept of "Lawfare." I objected to this term back when General Dunlap introduced it about 10 years ago. But it quickly picked up favor in the Justice Department. It implies that anyone who objects to the legality of a US action is engaged in an act of warfare.

Here is a quote from the "Lawfare Project" website:

> The enemies of the West and liberal democracies are pursuing a campaign of lawfare that complements terrorism and asymmetric warfare. Terrorists and their sympathizers understand that where they cannot win by advocating and exercising violence, they can attempt to undermine the willingness and capacity to fight them using legal means….
>
> The precedents set by lawfare actions threaten all liberal democracies equally. It is imperative that lawfare be opposed and that international human rights law and its interpretation be managed properly and in line with the tenets of democracy.[2]

It is certainly true that legal challenges can be frivolous, that allegations of violations by democratic governments can be fabricated. But it is also quite true that democratic governments, notably the United States, in the recent past have engaged in illegal detentions, interrogation, and surveillance—as well as questionable lethal drone attacks. If every allegation of wrongfulness by a democratic government were itself unlawful, then how would democratic institutions correct themselves under the rule of law?

The whole thing is quite reminiscent of the SLAPP (strategic lawsuits against public participation) controversy 30 years ago, in which industry would file suits against environmental groups[3] (who might then counterclaim for "abuse of process" or basic Rule 11). Those issues eventually just died away as people grew up and litigated under the rules.

This notion of "lawfare" seems to challenge the very idea of claims of abuse or human rights violations. So even drawing into question the validity of targeted killing could be considered lawfare—a chilling prospect in itself.

I think the ethics of propaganda allow for plenty of hyping and even misinformation but the degree of covert intervention into the internal affairs of another country has never been seriously delineated (how much did we do to foster Arab Spring?). If the topic is about cyber attacks, the ethical implications arise primarily from two points: the inability to control the weapon once it's loosed, and the degree to which you could bring down the infrastructure of another country and cause widespread suffering—a basic WMD.

In terms of international law, there has been a long-recognized conflict between the principles of self-determination and non-intervention. If an indigenous group is struggling to achieve independence or self-government, then their rights could include demands for assistance from outsiders, who are then subject to accusations of interference in the internal affairs of another sovereign.[4]

---

[2] http://www.thelawfareproject.org/what-is-lawfare.html (last visited Oct. 14, 2013).

[3] http://www.law.cornell.edu/wex/slapp_suit (last visited Oct. 14, 2013).

[4] The ability of an outside nation-state to come to the assistance of a rebel group traditionally depended on the fuzzy line between "insurgent" and "belligerent." But in recent times, a debate has arisen over whether it is permissible to intervene on behalf of liberation groups (see Gray 2000, pp 45–50).

The most important statement of the International Court of Justice on these matters is still Nicaragua v. United States:

> A prohibited intervention must [...] be one bearing on matters in which each State is permitted, by the principle of State sovereignty, to decide freely. One of these is the choice of a political, economic, social and cultural system, and the formulation of foreign policy. Intervention is wrongful when it uses methods of coercion in regard to such choices.

The implication of this statement is that methods other than coercion would not be wrongful. Indeed, the basic prerogatives of "sovereignty" on the part of the interfering state would seem to support its right to use not just "diplomacy" or "propaganda" but any means short of force in pursuit of its own foreign policy. Indeed, hamstringing other states from pursuing their aspirations of global governance could run counter to the very idea of sovereignty.

So far, so good, as concerns "informational" exchanges and even disinformation. But what about "material support" of rebel groups? In the Nicaragua case, the U.S. was found to have interfered in the affairs of another nation by training and equipping the Contra forces. But what if the U.S. had merely supplied money and organizational assistance without the training and equipment? Is this level of "informational warfare" prohibited?

It is hard to believe that support of rebels, short of supplying arms, would be found unlawful for two reasons. First, there is the very pragmatic difficulty of trying to prove the case. Any clandestine operative worthy of the name would cover his/her tracks sufficiently to avoid obvious involvement in rebel movements. Despite the well-publicized accusations of U.S. interference in the Egyptian uprising, nothing very concrete can easily be laid at the feet of the American government.

Second, substantively supplying organizational services is closer to supplying information than it is to supplying arms. Supplying organization and information are both protected elements of free speech in the American system, at least until one reaches the level of supplying "expert advice and assistance" as those terms are defined in the "material support" of terrorism statutes.

In Holder v. Humanitarian Law Project,[5] the Supreme Court upheld a prohibition on providing "expert advice and assistance"[6] on the ground that it was possible to distinguish between providing "advice or assistance derived from scientific, technical or other specialized knowledge" from those types of support that are more general in nature. The Court argued that the plaintiffs' professed desire to train members of a designated foreign terrorist organization (FTO) in peaceful dispute resolution could promote the illegitimate aims of the organization by "buying time to recover from short-term setbacks, lulling opponents into complacency, and ultimately preparing for renewed attacks."[7] In turn, the proposal to teach FTO members on how to petition bodies such as the UN for relief might yield monetary aid that could then be redirected to fund the organization's violent activities. Finally,

---

[5] 130 S.Ct. 2705 (2010).

[6] 18 U.S.C. § 2339A(b)(3).

[7] 130 S. Ct. at 2729.

the Court made quick work of the freedom of association. The statutory scheme was distinguishable from the Communist Party cases on the ground that membership was protected. "The statute does not penalize mere association with a foreign terrorist organization," only the provision of material support to FTOs. The Court asserted that the "statute does not prohibit being a member of one of the designated groups or vigorously promoting and supporting the political goals of the group…. [It] prohibits the act of material support."

Applying these thoughts in the context of a criminal prosecution for "material support" is slippery enough, but taking them into the international arena seems utterly inappropriate. The corollary of the U.S. individual freedom of speech for a nation is the right to pursue its foreign policy. To pursue that policy by providing support and assistance—not technical training nor tangible goods such as arms and equipment—is within the basic definition of the nation-state.

Did the U.S. violate international law by assisting the organizers of protests and demonstrations in El-Tahrir Square? It would be extremely difficult to make that argument.

But is it permissible to fund the political campaign of a favored candidate in an election? Surely not. Despite the U.S. Supreme Court's obstinacy in thinking that money is protected speech, it is surely an unwarranted interference in the affairs of another nation to fund political campaigns. In fact, U.S. law makes it a crime for a "foreign national to contribute to a campaign in the United States".[8] Money is not the same as information or assistance.

On the other hand, individuals are not the same as governments. If a U.S. employee or contractor in Iraq or Afghanistan campaigns personally for either the incumbent or an opposition candidate, it is difficult to say that there is a violation of international law—violation of U.S. employment contracts, perhaps, but not international law.

Providing money to a partisan campaign is more akin to providing arms and equipment to the rebels. Indeed, since money is fungible, the Supreme Court has recognized that it can be criminalized as "material support" to a terrorist organization. As a practical matter, there is nothing to prevent money provided for campaign purposes from being used for purchase of arms. Even on a theoretical or ethical basis, there is a world of difference between personal service and money, especially given the enormous disparity in wealth between the U.S. and some of the nations whose elections it might wish to influence.

"Informational warfare?" There is nothing wrong with putting out information, even disinformation. There is nothing wrong with providing personal assistance to groups organizing to promote self-determination. But there is something very wrong with providing campaign money to candidates in other countries.

Now to return to the distinctions with which I began, both money and information are different physical invasion of electronic infrastructure. A cyber attack on another country's banking, electrical, or other utilities systems would be the modern equivalent of "armed attack," which is permitted only for defensive purposes. In-

---

[8] 2 U.S.C. § 441e.

deed, many observers worry significantly about both cyber attacks and EMP attacks that can take out a country's infrastructure and cause massive loss. This is not an attack by of use of information, but it is an attack on information itself—in this instance, the information base on which a modern country operates.

In sum, the rules regarding information and warfare depend very much on what is being attacked, by what means, by whom. An individual can spend her money as she wishes, and a government can pursue its own informational policies. But a government is constrained not to interfere in the internal affairs of another country. Meanwhile, a cyber attack on infrastructure would be subject to the ordinary rules of the Law of Armed Conflict (LOAC) and its defensive postulates.

4.2 Normative Challenges to Cyber Warfare

In this Sect. I examine the normative challenges of cyber warfare through a critical review of the moral permissibility of preventive war. I claim that any advocacy of preventive intervention, however constrained, could gain undue legitimacy, leading to more war, not less. My claim is based on two factors—one, the slippery transition from preemption to prevention and the other, that even a limited provision of preventive war for justified self-defense, construed as a rare exception, can lead to a rather open-ended advocacy and use of it in the hands of a powerful state. In the case of the "Bush doctrine," we see a mix of both. Though couched in the language of preemption to make room for unilateralism in the guise of preemptive self-defense, the doctrine embraces far-fetched preventive measures. Accordingly, the issue is the moral permissibility of preventive war, regardless of its scope and the circumstances. Indeed, the provision of preventive war in self-defense can get unduly interventionist, making the world less secure. Today's scenario of covert information warfare accentuates this prospect.

The most pronounced instance of the blurring of the distinction bewteen preemption and prevention is found in the Bush doctrine of 2002, largely in response to the September 11, 2001, terrorist attacks on the United States. The broad mandate of the Bush doctrine effectively makes the idea of "global safekeeping" an important part of national security strategy, giving the United States an open-ended unilateral license to respond militarily, in the name of "war on terror," to any acts or events in the world based solely on the internal perception of the United States.

Though most contemporary political and legal theorists advocating preventive use of force find the Bush doctrine too broad, they feel compelled to respond to the challenges of the changing nature of warfare in the twenty-first century. Consequently, the traditional debates of just-war have become a major focus of controversy in these defining years of unconventional warfare. A similar major turn in rethinking the just-war concerns occurred during the Second World War where the distinction between combatants and non-combatants blurred, making that war the first truly "total war." It compelled the Allied forces to navigate across a moral divide in deciding whether to undertake massive bombing of German civilian targets for military and strategic reasons. "I see this idea of just killing civilians and targeting civilians as being unethical—though the most unethical act in World War II for the Allies would have been allowing themselves to lose," says military historian Conrad Crane, quoted in the 2010 PBS Television's American Experience segment titled "The Bombing of Germany." We find the echo of Crane's words in Michael Walzer's classic restatement of the just-war doctrine. He writes: "But if there was no other way of preventing a Nazi triumph, then the immorality [of creating massive terror by targeting the non-combatant] …was also, simultaneously, morally defensible" (Walzer 2004, pp. 34–35). For Walzer, in cases of "supreme emergency," rules of war can be breached "when we are face-to-face not merely with defeat but with a defeat likely to bring disaster to a political community" (Walzer 2006, p. 268).

The just-war dilemma of the Allied leaders over bombing the German civilians was prompted by the German bombers attacking London for 57 consecutive nights, which indicates that the Allied response was directed at a "face-to-face" situation of dire catastrophe. The quandary facing today's political theorists who draw from the just-war tradition is provoked by a new set of challenges unique to the new century. The understanding of a "face-to-face" danger in today's world could take a whole new meaning in view of the unconventional nature of warfare and the specter of WMD. The question now is not only justifying first strike but deciding on how much in advance of the perceived threat, given the potential for catastrophic consequences if the threat is given the time to be carried out. The certainty factor of an imminent danger debated by the just-war theorists in the sixteenth and the seventeenth centuries is now put to severe test in view of this new challenge.

The situation is especially made complicated in view of today's scenario of cyber warfare in which a technologically advanced nation may undertake riskless covert warfare to thwart a perceived danger in another country in the name of preventive intervention for self-defense. As discussed in Sect. I above, supplying information is different from supplying arms, though both can be aimed at changing the direction of a conflict. The latter is a violation of state sovereignty and thus prohibited under international law. But covert cyber warfare is a most eggregious form of interference in the internal affairs of a state since it has the potential of bringing down the infrastructure of a country, but international law lacks specific guidelines regarding the rules of such covert operations. The morality of such warfare is also faced with unresolved issues if the imperative of self-defense is brought in as a justification for such intervention. Preventive intervention is a murky issue in the just-war thinking, so just-war doctrine does not provide much moral clarity in this debate. We may need to look elsewhere for moral guidance on this matter.

The blurring of the distinction between preemption and prevention is at the heart of the issue here. Advocacy of preventive war for justified self-defense, even when construed as a rare exception, can be rather open ended and liable to be misused. We see this in the Bush doctrine's espousal of unilateralism in the name of self-defense couched in the language of preemption, though it embraces far-fetched preventive measures. Legitimizing principled preventive war, however constrained, can give a powerful nation the moral license to expand the principle by pushing it in the direction of its own convenience. Yet some prominent contemporary just-war theorists who reject the Bush doctrine's expansive and reckless interventionism nonetheless advocate a limited provision of preventive war, even unilateral if warranted, for justified self-defense in cases of dire necessity. Their concern is to stay within the spirit of international law and devise means of accountability in offensive wars, with the goal of finding ways to respond to the new threats to peace and security posed by unconventional warfare and unconventional weapons systems. They rightly note that unilateralism in preventive ventures based on subjective and open-ended assessment of security threats can go horribly wrong in its calculations of anticipatory events and developments, and because it lacks political legitimacy and legal authority, it sets a dangerous precedent. In contrast, their provision for preventive use of force is primarily multilateral, guided by a mix of the just-war criteria and legal propriety, putting emphasis on collaboration whenever possible and citing the UN Security Council as the venue for open arbitration and debate for procedural legitimacy (Doyle 2008, Luban 2004, Buchanan and Keohane 2004). Their guidelines for assessing the gravity of the situation requiring prevention display a judicious blend of substantive and procedural considerations, including such factors as severity of threat, the likelihood of its occurrence, just-war criteria of legitimacy, and the legality of the threat and the proposed response.

Nonetheless, these guidelines are open-ended and can be misused. Just-war legitimacy criteria such as proportionality, necessity, and last resort are matters of disputation and prone to subjective interpretation, especially if a go-alone provision is allowed in the guidelines. Indeed, the just-war doctrine's major flaw is that it allows self-interested interpretation by the contesting parties (Myers 1996). The assessment of severity and likelihood of threat in anticipatory circumstances is no less subjective and open to mistakes or abuse. And the idea of legality is a moot question in claims of existential threat. As Michael Walzer has famously stated: "necessity knows no rules" (Walzer 2006, p. 254). Thus, these guidelines leave open the possibility that a powerful nation with global hegemony can construe them as an open-ended license to respond militarily, in the name of self-defense, to any emerging or anticipatory events in the world based on its own perception.

The prospect of cyber warfare compounds this problem. The growing reliance in modern warfare on information and communication technologies makes the blurring of preemption and prevention all too likely, thus accentuating unresolved moral and legal dilemmas of preventive war. There are several reasons for it. Unlike conventional warfare, regardless of its sophistication, virtual war offers the prospect of being risk free, instant, covert, and causing no immediate combatant and non-combatant injury on the enemy side. Though virtual war has the potential of making the entire infrastructure of a country dysfunctional, thereby causing untold suffering, it is still considered "clean" because it does not directly target people.

In the increasingly escalating use of drone attacks in the name of just-war where drones are often termed "moral predators," thus making obligatory their uninhibited use, unresolved moral and legal questions abound. Though drones are unmanned military robots and exemplify the advanced sophistication of military technology, they are still a step away from the specter of virtual war. Even then, deployment of drones alters the reciprocal vulnerability of a conventional war and makes the asymmetries of power more pronounced by making military operations risk-free for the side using drones. This prospect has the likelihood of misuse of military options in the name of preventive intervention. In commenting on the frequent use of unmanned drones in today's US military combat overseas, Peter W. Singer writes:

> "And now we possess a technology that removes the last political barriers to war. The strongest appeal of unmanned systems is that we don't have to send someone's son or daughter into harm's way. But when politicians can avoid…the impact that military casualties have on voters and on the news media—they no longer treat the previously weighty matters of war and peace the same way." (*The New York Times*, January 22, 2012, Sunday Review)

In other words, risk-free combat technology can increase the likelihood of their use. In fact, one can make the more general claim that the permissibility of preventive use of force can make war all too tempting and frequent. This is especially true with the prospect of cyber warfare which is not only risk-free like drones, but also is instant, covert, and causes no immediate death on the other side. But all these features raise moral and legal conundrums. In essence, legitimizing preventive war, however constrained, can give a nation the moral license to expand the principle by pushing it in the direction of its own convenience. And if preventive war is made easy due to the use of cyber technology in military operations, then the chances are that much greater that the technology would be put to use in the name of preventive intervention. The certainty factor of an imminent danger, already compromised in the need for expanded preemption due to the presence of WMD in today's unconventional warfare, is now put to severe test in view of the new challenge of cyber warfare, making the claims of moral mandate in the slippery transition from preemption to prevention that much easier. But this trend is making the world progressively less secure. Neta Crawford's observation is worth noting here: "In sum, a preemptive-preventive doctrine moves us closer to a state of nature than a state of international law" (Crawford 2003).

The mindset of preventive war perpetuates the anxiety of living under the shadow of war, whereas "the stress of living in fear should be assuaged by true prevention—arms control, disarmament, negotiations, confidence-building measures, and the development of international law" (Crawford 2003, p. 36). These preventive measures are instances of proactive non-intervention that use the soft power of diplomacy and democratic collaboration. This may be a long and hard road that promises no quick results but, then, if we're looking for a fail-safe quick path to peace and security in today's murky and uncertain world, nothing can take us there. Preventive interventions make things only worse. We should pay heed to Grotius who said: "Human life exists under such conditions that complete security is never guaranteed to us."[9]

---

[9] Grotius (1625: 184), cited by Larry May in his chapter in Chatterjee (2013a). Portions of Section II are excerpted from Chatterjee (2013b).

References

Buchanan, Allen, and Robert Keohane. 2004. The preventive use of force: A cosmopolitan institutional approach. *Ethics and International Affairs* 17 (1): 1–18.

Chatterjee, Deen. 2013a. *The ethics of prventive war.* Cambridge University Press.

Chatterjee, Deen. 2013b. Enough about just war, what about just peace? The doctrine of preventive non-intervernation. In *The ethics of preventive war*, Hrsg. Deen Chatterjee. Cambridge University Press.

Gray, Christine. 2000. *International law and the use of force*. 45–50. Oxford.

Crawford, Neta. 2003. The slippery slope to preventive war. *Ethics and Interantional Affairs* 17 (1): 35.

Doyle, Michael. 2008. *Striking first: Preemption and prevention in international conflict.* Princeton: Princeton University Press.

Luban, David. 2004. Preventive war. *Philosophy and Public Affairs* 32 (3): 207–248.

Walzer, Michael. 2004. *Arguing about war.* New haven: Yale University Press.

Walzer, Michael. 2006. *Just and unjust wars: A moral argument with historical illustrations*. 4th ed. New York: Basic Books.

Myers, R. J. 1996. Notes on the just war theory: Whose justice, which wars? *Ethics and International Affairs* 10 (1), 115–130.

Part II
Just Information Warfare

Chapter 5
Permissible Preventive Cyberwar: Restricting Cyber Conflict to Justified Military Targets

George R. Lucas

Abstract Evaluations of cyber war and weapons range from denunciations of their widespread and indiscriminate destructiveness and deliberate targeting of civilian infrastructure, all the way to appraisals of cyber warfare as a morally preferable, less destructive alternative to conventional warfare. This chapter will bring some order to this chaos by distinguishing permissible from impermissible forms of cyber conflict, as well as distinguishing genuine "warfare" from large-scale criminal or terrorist enterprises. The chapter will criticize the lack of discrimination often encountered in the formulation of cyber strategy and development of cyber weapons, and argue in favor of international governance and guidance that (with reference to proportionality, discrimination, and the principle of last resort) restricts the use of cyber weapons to justified military targets, using Stuxnet as a recent case in point. In ethics, we can infer or derive operable constraints on, and guidelines for acceptable practice by examining instances of what all agree is either good or bad practice, just as in international law, we recognize the evolution of customary law through the accepted conduct of otherwise law-abiding states. Hence, I will argue that an act of cyberwarfare is permissible if it aims primarily at harming military (rather than civilian) infrastructure, degrades an adversary's ability to undertake highly destructive offensive operations, harms no civilians and/or destroys little or no civilian infrastructure in the process, and is undertaken as a "last resort" in the sense that all reasonable alternatives short of attack have been attempted to no avail, and further delay would only make the situation worse.

The original version of this paper was delivered at a UNESCO-sponsored conference on Cyber War and Ethics at the University of Hertfordshire (July 1, 2011): http://www.elac.ox.ac.uk/downloads/Permissible%20Preventive%20Cyberwar%20UNESCO%202011.pdf.

G. R. Lucas (✉)
Ethics and Public Policy, Naval Postgraduate School, Mile Drive, 225-17, Pacific Grove, CA 93950, USA
e-mail: grlucas@usna.edu

L. Floridi, M. Taddeo (eds.), *The Ethics of Information Warfare*, Law, Governance and Technology Series 14, DOI 10.1007/978-3-319-04135-3_5,

5.1 "Cyber Anxiety" and Threat Inflation

Ours is the age of "cyber anxiety." Pundits opine, especially in developed, highly industrialized countries, on the global vulnerabilities to cyber attacks or to acts of cyber terrorism. Cyber security, and fending off cyber crime, are a constant obsession and an ongoing concern. The potentially indiscriminate and uncontrollable aspects of cyber weapons, once unleashed in acts of terrorism or warfare, is the subject of grim and dark prognostication. In many respects, the fear of uncontrolled proliferation and widespread destruction from cyber warfare has come to occupy a place in the public mind very similar to the current fear of terrorist attacks, or even more to the threat of uncontrolled nuclear destruction that haunted public consciousness during the decades of the Cold War. The situation of the U.S. and its allies in Western Europe *vis a vis* potential adversaries (like China or the Russian Federation) has, indeed, been portrayed as analogous to the nuclear cold war: a proliferation and virtual "arms race" in the cyber arena, with only a presumed balance of destruction holding adversaries at bay.

Apart from the Convention on Cybercrime sponsored by the Council of Europe a decade ago (Budapest 2001), however, not much progress made in the field of governance: that is, on discussions of the most likely ethical constraints on cyber conflict, or on the content of feasible treaties, or the formulation of bright-line statutes in international humanitarian law, that might serve to limit or regulate some of the most fearful or destructive prospects attendant upon cyber weapons development or permissible cyber tactics and strategy. The most detailed treatment of the ethics of cyber warfare has been the analysis of philosopher Randall Dipert of the University of Buffalo, writing in the December, 2010 issue of the *Journal of Military Ethics* (Dipert 2010).[1] Dipert, laments the relative lack of attention given to the ethics of cyber war, and cites a modest body of prior work in this field undertaken largely by computer scientists and policy advocates: Martin Libicki (RAND Corporation 2009), Herbert Lin (AAS 2009), and two colleagues at the Naval Postgraduate School, computer science professor, Neil Rowe, and Professor and Chair of the Department of Defense Analysis, Dr. John Arquilla.

Rowe primarily discusses the status of cyber warfare and weapons with reference to current statues of the law of armed conflict, and complains quite appropriately that many of the strategies and weapons for cyber conflict currently under development constitute potential violations of prevailing international humanitarian law (LOAC), in that they deliberately target, and aim to inflict widespread damage and suffering, and even injury and death, on civilian personnel and infrastructure (Rowe 2007, 2008, 2010). John Arquilla, who coined the phrase "cyber warfare" itself while at the RAND Corporation (Arquilla and Ronfeldt 1993; Arquilla 2003),

[1] More recently, from the perspective of domestic and international law, I commend a keynote address by Steven Bradbury, former head of the Office of Legal Counsel in the U.S. Department of Justice, entitled "The Developing Legal Framework for Defensive and Offensive Cyber Operations," delivered at the annual Harvard National Security Symposium in 2011, devoted to "Cybersecurity: Law, Privacy, and Warfare in a Digital World" (Bradbury 2011; see also Goldsmith 2011).

also wrote what is likely the very first, and I would claim, still the most original and path-breaking article on "ethics and information warfare (Arquilla 1999)." Like Dipert, Arquilla discusses principally the ethical issues, as opposed to the legal status, of cyber conflict, and runs its principal strategies and tactics through the lens of just war theory. I will return to Arquilla's pioneering observations in my conclusion.

Dipert's more recent account of the ethics of cyber warfare, while certainly not the first, is surely the most complete and up to date ethical account from the standpoint of the current status of the technology of cyber conflict, and also the most thorough and fully informed analysis from the standpoint of philosophy, ethics, and particularly just war theory. In keeping with what many other analysts concluded over the past decade with respect to terrorism, counterinsurgency, and irregular warfare, Dipert concludes likewise in the cyber case that the tactics and weapons of cyber conflict are such as to render traditional law and morality obsolete, or at least, largely inapplicable. His overall conclusion is that cyber conflict is so utterly unlike conventional war, and its weapons and tactics so novel and unprecedented, that an entirely new regime of governance is called for. He thus echoes, and indeed, fans the flames of public anxiety over this mode of conflict.

For my part, I do not doubt the gravity of the threat (as Arquilla described it over a decade ago), nor do I dispute the seriousness of the concerns that Dipert now raises and discusses. I do think this topic presently suffers from a certain amount of hysteria. It is certainly true that cyber conflict is, like irregular warfare (IW) and terrorism, a substantial challenge to our conventional thinking about war and armed conflict, and will certainly call for some disciplined and careful analysis, and some constructive efforts to meet the challenge of effective governance in the near future. The fear that we might unwittingly or inadvertently unleash a widespread and unrestrained, and highly destructive conflict in the cyber arena, in particular, as an act of war is a very real concern.

But public discussions, including the essays I have cited, often fail to distinguish, or even attempt to distinguish with sufficient care, between different kinds of cyber conflict:

1. what might be called cyber *vandalism* (a hacker breaking into, and lurking in defense information systems);
2. acts of cyber *crime* (in which data are damaged or stolen, or services denied, for personal or corporate gain);
3. cyber espionage (what might be accurately described as acts of cyber vandalism and cyber crime carried out by states or commercial corporations;
4. cyber *terrorism* (in which all of the foregoing things, and also damage and destruction to physical infrastructure are inflicted by aggrieved non-state agents in order to sow fear and confusion, and inflict widespread physical suffering upon random victims); and
5. genuine acts of cyber *warfare*, in which the latter sorts of things (physical damage, causing death, destruction, and widespread physical suffering) are done deliberately, to specified adversaries, in pursuit of political objectives or conflict resolution by states, governments, and their military and intelligence forces.

In passing, let me quickly remark that I am of the opinion that *the threat of cyber terrorism, in particular, has been vastly overblown*. Unlike IW and conventional acts of terrorism generally, genuine cyber warfare turns out to be a very expensive, labor intensive, and therefore remain a highly state-centric enterprise. Terrorists can engage in vandalism and crime, and have used the internet to great advantage for the purposes of conventional propaganda and dis-information. But they cannot easily develop true cyber weapons, or engage in acts of cyber warfare—nor have they yet been detected as doing so, or even trying to do so. To be blunt: neither the 14-year old hacker in a next-door neighbor's upstairs bedroom, nor the two- or three-person al Qaeda cell holed up in some apartment in Hamburg, are going to bring down the Glen Canyon and Hoover Dams. And that offers occasion for modest hope. I will return to explain these assertions, and to examine Professor Arquilla's analysis of the ethical principles governing them, in conclusion.

5.2 Discerning Acceptable and Unacceptable Practices in Cyber Conflict

For the moment, I want to make the case that there are acceptable forms of cyber conflict and cyber warfare that can be justified from the standpoint of just war theory. Indeed, such cyber conflict (as Neil Rowe has allowed) may in some instances be preferable to conventional war, and even to alternative forms of conflict resolution (such as economic sanctions), if properly conducted. I also want to remark that our actual experience of cyber warfare to date (and there have been several military strikes by governments), has not been all that bad, and is likewise such as to offer hope that the worst fears regarding cyber conflict may be somewhat exaggerated. Indeed, I think it is possible on the basis of experience to distinguish between morally justified and unjustified forms of cyber conflict, and to discern, quite remarkably, that those cyber strikes that have been conducted within the current constraints of law and morality (e.g., with respect to the prevailing principles of the law of armed conflict) have also, to date, proven more effective than those that potentially represent the commission of war crimes.

Let me begin with the "bad" attacks. One important such attacks that we know of were presumably unleashed by the Russian Federation against nearby adversaries in Estonia (in April, 2007), and another, also presumably by Russia, in Georgia (in July 2008). The first instance was basically a "distributed denial of service" (DDOS), overwhelming and shutting down service in a sophisticated country dominated by paperless government and heavy reliance upon internet financial transactions. A DDOS attack began around 20 July 2008 in Georgia, when "botnets" from all over the world began blasting Georgian computer services and networks with enormous amounts of useless data, much of which was eventually traced back to the RBN (Russian Business Network), an organized crime unit of Russian mafia. This was a prelude to conventional bombing and perhaps also intended as a prelude to full scale cyber-war (that was not carried through). The attribution of cyber attacks is, of course, a well-known problem, and no official source in Russia has ever admitted complicity in either case. What is significant, however, is that the first strike was an act of largely unprovoked aggression, that was apparently a response to actions by the eventual victim state that did not begin to rise to the accepted level of *causus belli* in international law. Also, and even more significantly, the first strike, far more than the second, relied almost exclusively on targeting civilians and civilian infrastructure. Happily, in neither case was extensive damage done, nor injuries sustained, nor were lives directly lost as a result of the use of cyber weapons.

It is difficult to gauge the effectiveness of the Estonian attack: certainly the cyber attacks seemed to be followed by a reduction or diminution of the most extreme forms of anti-Russian political behavior in Estonia that presumably had provoked them. The Georgian attacks, by contrast, seem to have constituted more a prelude or warm-up for conventional armed intervention: initially) (but mistakenly thought to be the first time that a conventional attack was deliberately preceded by a cyber attack [see Syria (2007) below] and apparently served to prepare the way for Russia's subsequent conventional armed intervention in Ossetia. Both attacks, from a political perspective, caused a great deal of resentment, and inflamed hostilities, making a political solution to either conflict relatively unlikely. Estonia requested at the time that NATO recognize a violation of sovereignty, so as to trigger the collective self-defense provision of the NATO treaty. Interestingly, that suggestion was rejected at the time on the grounds that "a cyber attack is not a clear military action (Schaap 2009)." In the second case, a preparatory cyber attack may have aided the success of the conventional intervention and occupation. But in neither of these known cases did the cyber strategy address, alter, or otherwise remedy or resolve the underlying political conflict.

Israel apparently likewise preceded its air strikes against a Syrian nuclear site at Dayr az-Zawr in September of 2007 with a full-scale cyber attack, though once again the details are murky, and formal attribution has never been made or acknowledged. In this case, far more so than in either of the above attacks, however, the preemptive cyber strikes were directed entirely against military targets: radar and air defense systems, much as a conventional attack might have been. Unlike the conventional case, however, the cyber attack attained the military objective of rendering defensive forces helpless, without widespread destruction of property or loss of life on either side resulting from the use of cyber weapons. Hence the cyber attack replicated the effects of a conventional armed attack, but achieved its objective with far less destruction, risk, or loss of life than would have accompanied a conventional attack designed to achieve the same purpose.

5.3 Cyber War and Just War

These three cases together offer an important set of evaluations that I want to take up with respect to some of the core criteria of just war theory: "just cause" and "last resort" with respect to the justification of war (*jus ad bellum*); and "proportionality" and "discrimination" (or, in international law, the principle of "distinction") with respect to the conduct of hostilities and specific applications of force (*jus in bello*). From the perspective of *jus ad bellum*, I would like to argue that the (presumed) Russian attacks, but especially the first attack against Estonia, lacked a sufficient just cause and were not undertaken in any meaningful sense as a last resort. Moreover, from the perspective of the just conduct of hostilities (*jus in bello*), of the two Russian attacks, the first was utterly indiscriminate, and was likewise disproportionate in its threat of harm, at least, when compared either to the harm Russia itself allegedly had suffered from Estonia civic policy, or any legitimate military objective that might have otherwise been under consideration. These observations and judgments are hardly surprising: the Russians have a long history of making too ready, indiscriminate, and disproportionate resort to force even when they have a legitimate objective whether in domestic or international situations. By comparison, interestingly, consider the (presumed) Israeli preemptive military cyber attack on Syria, preceding its conventional strike against their nuclear facilities. A conventional strike had been continuously threatened through diplomatic channels in the event of the Syrian government under Bashir al Assad ever attempting to develop a nuclear weapons program. There was arguably adequate justification leading up to a conventional attack on the illegal nuclear facilities Syria was attempting to construct, and thus also justification for the preparatory cyber attack to disable Syrian air defense systems. Importantly, both the cyber and conventional military actions were undertaken only after reasonable diplomatic efforts had failed. The targets of cyber strikes were entirely military, and the overall damage inflicted as a result rather minimal, and arguably proportional to the harm threatened, the wrong done, and the military objective in question.

If this assessment is correct, it suggests (in marked contrast to Dipert's conclusions) that not all cyber conflict escapes the analytical framework of classical or conventional just war theory, and vice versa, that consideration of just war doctrine may effectively guide the conduct of cyber war, even as it attempts to do for conventional and irregular warfare. In the latter case, one of the most controversial topics in the past decade has been the justification of preventive war, undertaken against an enemy who has, as yet, done no actual harm, but represents a future threat of harm. Classical just war doctrine rejects the legitimacy of a cause for war that does not involve the actual (rather than merely threatened) infliction of harm through an act of aggression. And yet this has not seemed to many recent analysts (myself included) to address adequately the dilemma of, e.g., the menace of rogue states, or terrorist preparations for attacks that have the aspects of a criminal conspiracy not yet fully consummated.

In this regard, it is instructive to consider a fourth, more recent case: that of Stuxnet, which the *New York Times* in January of 2011 described as "the most sophisticated cyber weapon ever deployed (William et al. 2011)." Once again, the problem of attribution is vexed: no nation or coalition has come forward to claim credit, or accept blame, for having engaged in what has gradually come to be identified as an act of preventive warfare (Gross 2011). Suspicion falls heavily on those who stood to gain the most from the attack, and perhaps on those who smile the most broadly, without comment, when the event is cited. The details, such as are known, are likely familiar, so a brief summary of the key points of this act of war will likely suffice for the purposes of this essay.

The Stuxnet virus is a cyber "worm" of unknown origin, apparently developed and released in a number of countries in 2009 (one analysis, by the Symanatec Corporation, postulates that the worm was initially released in Indonesia, rather than, as subsequently alleged by a secret agent in Iran itself). By July 2010, the Stuxnet worm was known to have infected computers all over the world. Nearly 60 % of infected systems were located in Iran (although others ranged from India, Pakistan, Indonesia and Azerbijan to the U.S. and Europe), and so after some initial confusion, Stuxnet was assumed to be a cyber weapon targeted at Iran, that had subsequently failed in its primary purpose and run amok, spreading uncontrollably to unintended targets all over the world, and thus demonstrating how indiscriminate and destructive cyber weapons were likely to be. This was the assessment of Stuxnet offered in a footnote in Professor Dipert's essay (Dipert 2010, p. 407, n. 3).

A study of the spread of Stuxnet by Symantec showed that the main affected countries in the early days of the infection were Iran, Indonesia and India:

Country	Infected computers (%)
Iran	58.85
Indonesia	18.22
India	8.31
Azerbaijan	2.57
United States	1.56
Pakistan	1.28
Others	9.2

What was a reasonable assessment at the time, however, turned out to be substantially incorrect. Unlike most malware, Stuxnet did little harm to computers and networks that do not meet specific configuration requirements. "The attackers took great care to make sure that only their designated targets were hit…It was a marksman's job."[2] While the worm is promiscuous, it renders itself inert if Siemens software is not found on infected computers, and contains safeguards to prevent each infected computer from spreading the worm to more than three others. All copies of the virus are set to erase themselves on 24 June 2012 (William et al. 2011).

Why Siemens software? The virus attacks and destroys nuclear centrifuges manufactured by Siemens, overriding the proprietary software and overloading the centrifuges themselves until they self-destruct. It does so cleverly, in the manner of the Hollywood film, *Ocean's Thirteen*, by running a second sub-routine (known as a "man in the middle") that disguised the damage in progress from operators and overseers until too late to reverse. One line of code restricts this damage, however, only to an array or "cascade" of centrifuges of a specific size (exactly 984). In sum, unless you happen to be running a large array of Siemens centrifuges simultaneously, you have nothing to fear from this worm. It is an extremely sophisti-

[2] Comment of Ralph Langner, a computer security expert in Hamburg, Germany, quoted in *NY Times* (William et al. 2011).

cated weapon: estimates are that it must have been years in the development, with large teams of experts and access to highly restricted and classified information and equipment. This is not something a terrorist group, or even likely a well-organized and funded criminal organization could have undertaken (and certainly not a single 14 year-old hacker!). The investment of time and resources and expertise were simply beyond any but a well-positioned state or coalition to effect. The damage was done exclusively to a cascade of centrifuges, illegally obtained and operated in an otherwise hardened and highly protected site at Natanz, in Iran, in explicit violation of the nuclear non-proliferation treaty. The damage sustained within Iran to its clandestine and internationally-denounced nuclear program was gauged at the time as being substantial, putting back its weapons development program by several years, at least.

In comparison with the previous cyber conflicts cited above: there was a good and justifiable reason, reluctantly sanctioned in the international community, to undertake military action against Iran's nuclear weapons program. Famously, diplomatic efforts and other, non-military measures had been undertaken for years without success. The harm or risk posed is extremely serious, but it is future harm: i.e., harm threatened, rather than inflicted, so Stuxnet was clearly a preventive (preemptive?) attack. The target was wholly military, and damage confined to the targets identified. There was no collateral damage of any meaningful or significant sort to lives or property: civilian personnel and infrastructure were apparently neither targeted nor affected. Most importantly, when compared against "Operation Babylon," the conventional Israeli air raid against Iraq's nuclear program at Osirik on June 7, 1981, this cyber strike involved far less damage, harm, and risk of either for all concerned.

Still there are concerns raised that the promiscuous spread of the worm has now made this destructive weapon available to users all over the world, who might tweak it and release another verions.[3] This concern about Stuxnet as an "open-source weapon" available for downloading by anyone, however, demonstrates a widespread and fundamental misunderstanding of the nature of cyber "weapons," to which Neil Rowe has called attention in his work. They are not like nuclear warheads or RPGs, simply obtainable and re-useable by anyone. Rather, they are by and large "one off:" once a given cyber weapon has been used, that is to say, its function is revealed, and anti-virus and security protections are quickly developed and disseminated against it, and the original weapon is seldom if ever itself reused, or usefully replicated. While "old generations" of computer viruses "hang around" and sometimes infect woefully under-securitized computer systems, there is as yet no known instance of a truly effective cyber weapon ever having been reversed engineered and re-used in an effective attack.

[3] This concern is voiced explicitly in the online "infographic" documentary, Clair (2011).

5.4 Establishing Norms for Ethical Cyber Conflict

So now let me return in conclusion, as promised, to the work of John Arquilla. Interestingly, in his path-breaking article, "The Ethics of Information Warfare" (1999) over a decade ago, Arquilla outlined what I take to be an argument for permissible preventive cyber attack. Though obviously not as familiar with the broader range of classical JW doctrine as Dipert and subsequent just war/ethics experts, Arquilla nonetheless homed in on precisely the most relevant features of morally-justified conflict: a grave and morally sufficient reason or just cause for war, a record of prior good faith attempts to resolve the conflict short of armed attack that made such war a necessary "last resort," and, in the targeting and tactics, a focus solely on threatening and strategic military targets, with the likely prospect of confining harm almost entirely to those targets, and entailing no risk to, let alone deliberate targeting of, civilian personnel or infrastructure. Under such severe constraints, Arquilla concluded, a cyber strike might be morally justified (*supra*, n. 7: 392–393). And I would add: such an attack would appear to be morally justified by such considerations, *even when it might otherwise constitute a preventive attack.*

Stuxnet conformed almost perfectly to Arquilla's constraints, so closely as to raise a kind of suspicion that its perpetrators had read his article, and followed his own outline of the relevant moral constraints virtually to the letter. For my part, I'm inclined to agree that the circumstances warranted such a preemptive attack, and that, as designed and carried out, Stuxnet was an effective and morally justified military cyber attack. It shows that cyber war can be an effective alternative to conventional war, when less drastic forms of conflict resolution have been tried in good faith, and have failed. And, contrary to the fears of Dipert and others, such weapons and tactics can be designed to be effective, discriminate, and to inflict proportionate damage on their targets—far more so than conventional attacks.

Finally, I mentioned in passing above that this sophisticated weapon, and effective cyber weapons and strategy generally, were still expensive, skilled, labor-intensive, and therefore state-centric enterprises. No terrorist could, nor has, attempted anything like this. An effective weapon of cyber warfare like Stuxnet, at least at present, simply outstrips the intellectual, organizational, and personnel capacities of even the most well-funded and well-organized terrorist organization, as well as those of even the most sophisticated international criminal enterprises. If one is going to bring down hydro-electric generators, nuclear centrifuges, and air traffic control, then one needs direct access to such devices or systems and the software that operates them, as well as an intimate knowledge of their operations. The 14-year old neighbor, in particular, who skipped (and subsequently flunked) physics and engineering classes to concentrate on his social networking skills lacks the requisite knowledge, as well as the access to the relevant hardware. If he succeeds in hacking into a defense department computer, he won't have a clue of what to do there (other than to carry out the cyber equivalent of spray-painting artistic graffiti on subway cars). Centrifuges and hydro-electric generators, for their part, do not fit neatly into terrorist apartments in Hamburg, or sadly, even into the most well-equipped public high school laboratory. Admittedly, the air traffic control scenario presents a more ominous threat, from the standpoint of both terrorism and even vandalism or "hactivism," but would still require at minimum the leadership or assistance of a disaffected air traffic controller with years of experience and extraordinary security clearances.

That is moderately encouraging news. In addition, I believe our experience of states as entities with political interests, unlike the usual case of terrorists and non-state actors, makes these activities amenable to good governance. In the Stuxnet case, we have an example of what good governance might well license. In the other instances, we have examples of less justifiable actions (such as the indiscriminate and wonton targeting of civilians and civilian infrastructure) that might reasonably be renounced by all sides, without any discernable loss of political advantage. Rowe, for example, has suggested a plausible procedure for attribution in a crisis that, like the nuclear-era red-phone "hot line" between Washington and Moscow, would help avoid precipitating a kinetic response in the case of mistaken suspicion. Arquilla (1999, p. 396) recommended adopting a "declaratory doctrine of 'no first use' of information warfare against largely civilian targets." This simple step addresses a principal concern of cyber critics, like Dipert and Rowe, that the core strategies of cyber war and weapons are premised on the illegal targeting of civilians and civilian infrastructure,[4] while still allowing for strikes against military targets (operations centers, logistics, and command and control nodes). And, he adds, this policy still allows for retaliatory strikes in the event that one's own civilian targets are attacked. Finally, legal expert Stephen Bradbury, while echoing the current U.S. opposition to any new international conventions or cyber arms control agreements, argues that the accepted norms and limitations in the cyber arena will develop through the practice of leading nations restricting their behavior in conformance with the established rules and customs of warfare.

It is time to acknowledge what we have now discerned through such practices, good and bad, to move ahead with such discussions and the formulations of relevant treaties and protocols, and to put to rest some of the more extreme, hysterical, and unfounded fears about cyber conflict. Outlawing indiscriminate destruction, and deliberate civilian targeting, constitute a good beginning, and these cases show that such measures would not rob states of their abilities to conduct political conflict effectively within the accepted bounds of law and morality.

[4] I have argued elsewhere ("Postmodern War," *Journal of Military Ethics* 9, no. 4 (2010): 296) that this tendency to target civilians in cyber conflict stems from the overwhelming influence of intelligence and espionage, or clandestine services communities in the formulation of strategy and development of weapons, as contrasted with the conventional war-fighting community (even though a preponderance of the participants, from General Keith Alexander and VADM William McCollough on down, wear (or wore) military uniforms). In espionage, covert action, and "psych ops," there is no restriction on targeting civilians (although this has begun to be questioned in the intelligence community's own discussions of professional ethics): See also Lucas 2013.

References

Arquilla, John, und David Ronfeldt. 1993. Cyberwar is coming! *Comparative Strategy* 12 (2): 141–165.

Arquilla, John. 1999. Ethics and information warfare. In *The changing role of information in warfare,* eds. Andy Marshall, et al., 379–401. Santa Monica: RAND Corporation.

Arquilla, John. 2003. Interview for PBS “Frontline”. http://www.pbs.org/wgbh/pages/frontline/shows/cyberwar/interviews/arquilla.html. Accessed 4 March 2003.

Bradbury, Steven G. 2011. Cybersecurity: Law, privacy, and warfare in a digital world. Harvard National Security Law Review Annual Conference. http://harvardnsj.org/wp-content/uploads/2011/02/Vol.-2_Bradbury_Final.pdf. Accessed 4 March 2011.

Budapest. 2001. Council of Europe, Convention on cybercrime. http://conventions.coe.int/Treaty/EN/Treaties/html/185.htm. Accessed 23 Nov 2001.

Clair, Patrick. 2011. Stuxnet: Anatomy of a computer virus. http://vimeo.com/25118844.

Davis, Joshua. 2007. Hackers Take Down the Most Wired Country in Europe. *Wired Magazine* 15 (9): http://www.wired.com/politics/security/magazine/15-09/ff_estonia?currentPage=all. Accessed 12.31.2012.

Dipert, Randall R. 2010. The ethics of cyberwarfare. *Journal of Military Ethics* 9 (4): 384–410 (December 2010).

Goldsmith, Jack. 2011. Steven Bradbury on cyber security. In the Blog, Lawfare. http://www.lawfareblog.com/2011/04/steven-bradbury-on-cybersecurity/. Accessed 18 April 2011.

Gross, Michael J. 2011. A declaration of cyber-war. *Vanity Fair*. Condé Nast. http://www.vanityfair.com/culture/features/2011/04/stuxnet-201104. Accessed 3 March 2011.

Libicki, Martin C. 2009. *Cyberdeterrence and cyberwar*. Santa Monica: Rand Corporation.

Lin, Herbert S., et al. 2009. *Technology, policy, law, and ethics regarding U.S. Acquisition and use of cyberattack capabilities*. Washington, DC: National Research Council/American Academy of Sciences.

Lucas, G. R. Jr. 2013. *jus in silico*: moral restrictions on the use of cyber warfare. In *The Routledge Handbook of War and Ethics,* eds. Allhoff Evans, and G. Henschke Nicholas, 367–380. London: Routledge.

Rowe, Neil C. 2007. War crimes from cyberweapons. *Journal of Information Warfare* 6 (3): 15–25.

Rowe, Neil C. 2008. Ethics of cyber war attacks. In *Cyber warfare and cyber terrorism,* eds. Lech J. Janczewski, and Andrew M. Colarik. Hershey: Information Science Reference.

Rowe, Neil C. 2010. The ethics of cyberweapons in warfare. *International Journal of Cyberethics* 1:1.

Schaap, Arie J. Major. 2009. Cyber warfare operations: Development and use under international law. *Air Force Law Review* 64 (121): 144–145.

William, J. Borad, Markoff John, E. Sanger und David. 2011. Israeli test on worm called crucial in Iran nuclear delay. New York Times. http://www.nytimes.com/2011/01/16/world/middleeast/16stuxnet.html?_r=1. Accessed 15 Jan 2011.

Chapter 6
Moral Cyber Weapons

Dorothy E. Denning and Bradley J. Strawser

Abstract This paper examines the morality of cyber weapons, offering conditions under which they are not only ethical under just war theory, but morally preferred over their kinetic counterparts. When these conditions are satisfied, states not only have the option of using cyber weapons, but could even acquire a moral duty to do so over other forms of warfare. In particular, we show that states are morally obliged to use cyber weapons instead of kinetic weapons when they can be deployed for a purpose already deemed just under the law of armed conflict and without any significant loss of capability. The reason behind this moral obligation is that cyber weapons can reduce both the risk to one's own (putatively just) military and the harm to one's adversary and non-combatants. The paper discusses this obligation, using examples to illustrate cases where it does or does not apply. It also addresses several objections that have been raised about the use of cyber weapons, showing that they fail to fully counter the obligation to use cyber weapons derived from their reduction of risk and harm properties.

6.1 Introduction

The formation of military cyber forces in the United States, China, and other nations has stimulated considerable interest in topics relating to the deployment of cyber weapons in state-level conflicts. One area of particular interest, and the topic of this paper, concerns the ethics of using cyber weapons. So far, most scholarly attention

---

Approved for public release; distribution is unlimited. The views expressed in this document are those of the authors and do not reflect the official policy or position of the Department of Defense or the U.S. Government.

D. E. Denning (✉) · B. J. Strawser
Department of Defense Analysis, Naval Postgraduate School, Monterey, USA
e-mail: dedennin@nps.edu

B. J. Strawser
e-mail: bjstraws@nps.edu

L. Floridi, M. Taddeo (eds.), *The Ethics of Information Warfare,*
Law, Governance and Technology Series 14, DOI 10.1007/978-3-319-04135-3_6,

has focused primarily on whether cyber-attacks are legally permissible under the international law of armed conflict (LOAC) (Denning 2008; DoD 1999; Owens et al. 2009; Schmitt 1999, 2010; Wingfield 2000, 2009). LOAC derives from the just war theory tradition and consists of two primary divisions: *jus ad bellum*, the ethical justification for going to war, and *jus in bello*, the moral principles governing conduct within war. Both are concerned with state use of force, particularly armed forces, but the former specifies *when* that force may be applied, whereas the latter specifies ground rules for *how* it should be applied as part of the prosecution of a justified war. Together, they enshrine widely accepted ethical principles that are intended to promote peace and minimize the adverse effects of war on the world; the just war convention can be understood as a restraining influence on the moral horrors of war. However, these just war theory principles governing LOAC come from an era that predates cyberspace, leaving their applicability to cyber weapons a question open to interpretation. Some scholars have argued that cyberwarfare is so divergent from traditional forms of warfare that the principles of just war theory simply do not apply in any straightforward manner to cyber weapons (Dipert 2010). We disagree. We argue that, at least with some kinds of cyber weapons, not only can they adhere to the principles of just war theory but that a positive duty to employ them can arise, at least in certain contexts.

The main principles of *jus ad bellum* are codified in the Charter of the United Nations, which specifies the conditions under which member states may apply force against other states. The most relevant parts of the Charter include Article 2(4), which prohibits states from using force against other states during peacetime; Article 39, which gives the U.N. Security Council responsibility for responding to threats and acts of aggression; and Article 51, which gives states a right to self-defense. With respect to cyber weapons, the primary question has been whether cyber-attacks constitute the use of force and, therefore, fall under the provisions of the above articles. Two general approaches to force analysis have been proposed: equivalent effects analysis (DoD 1999) and the Schmitt analysis (Schmitt 1999, 2010). Equivalent effects analysis considers a cyber-attack to be a use of force if its effects are equivalent to those of an armed attack, while the Schmitt analysis uses a broad set of criteria to distinguish the application of armed force from permissible actions such as trade sanctions. Both can be difficult to apply to cyber-attacks, leaving considerable uncertainty as to whether a particular cyber-attack constitutes an illegal use of force under Article 2(4) or even whether such attacks could even be properly understood as acts of war (Rid 2011).[1]

Jus in bello principles are concerned with whether operations conducted during a state of war follow the moral principles of necessity, distinction, proportionality, neutrality, perfidy, discrimination, and superfluous injury. Applying these principles

[1] Rid (2011) has recently given a philosophical argument (rather than a legal argument) that the use of cyber weapons cannot and will not constitute war. He makes this case based on an analysis of the definition of war such that any given attack must be a lethal, instrumental, and political act in order to constitute war. And he finds cyber-attacks, in isolation, would not constitute all three criteria. We set Rid's analysis aside for the purposes of this paper.

to cyber operations has been less problematic, and many believe that cyber weapons could be employed ethically in the context of an otherwise just war (DoD 1999).

Rather than considering the question of whether cyber weapons can be used ethically under LOAC, this paper goes further and argues that, under certain conditions, their use can actually become morally obligatory. When these conditions are satisfied, states not only have the morally permissible option of using cyber weapons, but a moral duty to do so. In particular, we show that states are morally obliged to use cyber weapons *in place of* kinetic weapons for a just attack whenever doing so does not result in a significant loss of capability. The reason for this moral obligation is that cyber weapons reduce both the risk to one's own (putatively just) military and the harm to one's adversary and non-combatants. Overall, cyber weapons are more humane, less destructive, and less risky than kinetic weapons for achieving certain military effects.

The scope of our ethical analysis in this paper is highly constrained. It does not address the larger question of whether an arbitrary cyber-attack is ever permissible or whether such an attack should even constitute a use of force.[2] Instead, we restrict attention to cyber-attacks that are viable alternatives to kinetic attacks that have already been determined to be just under LOAC, given certain criteria. Arguably, if the use of certain kinetic weapons in a specific context is deemed morally permissible, then the use of cyber weapons to achieve all or some of the same effects should likewise be morally permissible. However, we go further than simply saying that deploying cyber weapons in that context is a morally acceptable alternative. We claim that there is a moral *duty* to use the cyber weapons under such circumstances. This moral obligation arises because cyber weapons can, in some cases, incur less risk and result in less harm than their physical counterparts, while still meeting the same level of mission capability of said physical weapons.

We will first discuss the general claim. Then we will discuss some objections that have been raised to the general idea of using cyber-attacks for military objectives that would function as objections to our argument for the moral obligation to use cyber-weapons. Ultimately, we find that the objections, while substantial, fail to fully counter the obligation to use cyber weapons derived from their reduction of risk and harm properties. Our overall approach follows that used by one of us, Strawser, to argue for a moral duty to employ unmanned aerial vehicles (UAVs) (Strawser 2010). This paper draws on Strawser's theory and approach in that work. However, whereas Strawser previously focused exclusively on risk, we here also consider harm.

---

[2] See Dipert (2010, p. 393) for an argument that the preemptive use of cyber weapons would likely be morally preferable over a similar preemptive kinetic attack, and possibly even morally permissible, precisely for the kinds of advantages cyber weapons have that we rely on in this paper (i.e. their general non-lethality and lesser degree of destruction). On the use of force question, see Rid (2011) noted above.

6.2 Cyber Weapons as Ethically Obligatory

Strawser argues for the use of UAVs from the principal of unnecessary risk (PUR), which he formulates as follows:

> PUR: If X gives Y an order to accomplish good goal G, then X has an obligation, other things being equal, to choose a means to accomplish G that does not violate the demands of justice, make the world worse, or expose Y to potentially lethal risk unless incurring such risk aids in the accomplishment of G in some way that cannot be gained via less risky means.

An important aspect of this formulation is that Y is being ordered to accomplish goal G, which itself is a good and fully justified goal worthy of pursuit. The principle does not prohibit Y from choosing a risky approach on his or her own accord. Rather, it only prohibits someone from ordering Y to use an approach than carries with it avoidable risks.

Taking this principle to be uncontroversial,[3] Strawser applies it to the military use of UAVs with the following operating principle (OP):

> OP: For any just action taken by a given military, if it is possible for the military to use UAV platforms in place of inhabited aerial vehicles without a significant loss of capability, then that military has an ethical obligation to do so.

Strawser argues for OP on the grounds that because UAVs do not risk the lives of their remote pilots, militaries are obliged to use them in circumstances where they do not result in a loss of capability to conduct an operation that is otherwise deemed just in accordance with the principles of *jus ad bellum* and *jus in bello*. In short, in the context of a fully justified war effort, it would be wrong for a military commander to order a manned aircraft operation when the same result can be achieved with an unmanned one and doing so does not in any way worsen the warrior's ability to fight justly. This is because to do otherwise would place an unnecessary risk on the warfighter. Strawser also observers that because unmanned missions are generally less costly than manned ones, this could provide further grounds for a moral obligation for militaries to use them, as money is then freed up for more worthy social goals. However, because PUR provides a less contingent and ultimately more normatively compelling reason for using UAVs, he uses it for the moral obligation's derivation.

Turning now to the use of cyber weapons, we first observe that, like UAV strikes, cyber-attacks can be launched and conducted remotely, making them less risky to military personnel than when engaging in kinetic strikes. Thus, the principle of PUR applies to cyber-attacks for much the same reason as it would for UAVs. However, certain kinds of cyber-attacks potentially have a further moral advantage in that a given military objective may be achievable without causing any loss of life or

[3] Although we cannot pursue it here, not all have found even this modest principle uncontroversial. Uwe Steinhoff (forthcoming) argues that it is, indeed, controversial and that acceptance of it could involve several normative problems. Steinhoff's objections to the PUR deserve response, but such a discussion lies outside of the scope of this paper.

physical damage to the adversary or innocent third-parties and noncombatants. That is, cyber weapons could cause considerably less harm than the kinetic weapons they replace, while still accomplishing a justified military objective equally as effectively. Thus, there are two morally compelling reasons to use cyber weapons in place of physical weapons where possible: they can reduce both the risk to one's own (presumably just) military forces and they can reduce the harm incurred to the adversary and others.

As with UAVs, we formulate the moral obligation to use cyber weapons first in terms of a general principle, which now factors in both risk and harm. Calling it the Principle of Unnecessary Risk and Harm (PURH), it states:

> (PURH): If X gives Y an order to accomplish good goal G, then X has an obligation, other things being equal, to choose a means to accomplish G that does not violate the demands of justice, or cause unnecessary harm and incur unnecessary risk, unless incurring such risk or delivering such harm aids in the accomplishment of G in some way that cannot be gained via less risky or less harmful means.

The principle of PURH is nearly identical to PUR, except for the addition of harm as a disvalue to avoid in the ordering of just agents in pursuit of a good goal. The duty to not "cause unnecessary harm" could be seen as derived from the PUR's original demand to not "make the world worse." But the PURH aims to make the avoidance of unnecessary harm an explicit part of the formulation. Two central components of traditional *jus in bello* principles are that of proportionality and military necessity. Combined, these principles are generally taken to mean that just forces ought to use as minimal force and deliver as minimal harm as is militarily necessary for, and proportionate to, accomplishing a given just objective. Military operations in pursuit of a just cause generally result in some harm, at least to the adversary. Yet the just war tradition demands that we avoid unnecessary harm to the extent possible in the prosecution of a just war. If some of the harm of war may be avoidable by using cyber weapons instead of kinetic weapons, it seems that the just war tradition would demand that we so use cyber-weapons, where possible.

A brief discussion on this aspect of the proportionality and necessity constraint on just action within war is worth briefly exploring. Usually the proportionality constraint (*in bello*) works to limit the amount of destructive force permissible to use for a given attack such that the predicted damage done is proportionate to the relative importance of given objective. That is, that the damage done is "worth" the given objective and not excessive compared to the military advantage gained by the attack. Here the relative import of a given objective will be tied directly to the good it does towards the just cause. For example, using a nuclear bomb to take out one lone enemy soldier (which presumably would do very little to advance the success of the just cause but would cause tremendous collateral damage) would be grossly disproportionate.

But proportionality is very closely linked to our understanding of the *jus in bello* principle of military necessity. Necessity can be seen as a growth out of the restraint implicated by proportionality. It demands not only that the force used is a proper "fit" to the objective, but also that only the minimum amount of force necessary to accomplish a mission is used for any given objective.

That means that warriors fighting on behalf of a just cause ("just soldiers") will bear *different* duties vis-à-vis the strictures of proportionality and necessity depending on the options available to them. Say a group of just soldiers, W, is engaging a set of enemy (and putatively unjust) soldiers, Z. W has available to them three means (A, B, or C) of attacking Z, each of which would be equally effective at meeting the given mission objectives W seeks on behalf of the just cause. A is a large bomb that would obliterate Z, but will also destroy the building Z occupies, cause massive damage to the surrounding country-side such as burning up agricultural fields, destroy other nearby buildings, (unintentionally) kill some nearby noncombatants, and so on. B is a bomb similar to A, but its blast radius will not extend much beyond Z, although it will destroy the building Z is occupying. C is a weapon which will target the individual members of Z, but will not do any damage to the building they occupy or surrounding area whatsoever. Recall: W has high confidence that A, B, and C are each equally likely to succeed and they have equal access to all three choices.[4]

Under such a scenario, proportionality demands that W use C over A and B if they are to be in alignment with *jus in bello* principles. Were W to use A or B in this case, they would be in violation of *jus in bello* and would not be acting justly in war; they have a moral *obligation* to use C and an obligation to *not* use A and B. But if W found themselves in a scenario whereby they *only* had access to A or B, then the obligation against using A and B would not obtain. In that case, of course, they would be obligated to use B, and obligated to not use A. The point is that proportionality and necessity restrict just actors to use only as much force (resulting in as much harm and risk) as is required to accomplish a given act.[5]

Like PUR, we consider PURH to be relatively uncontroversial and, within the context of war, it follows from the strictures of proportionality, as just shown.[6] If one need not incur unnecessary risk or harm to carry out a just act, one should not. Applying PURH to cyber weapons leads to the following cyber operating principle (CyberOP):

> CyberOP: For any just action taken by a given military, if it is possible for the military to deploy remote cyber-attacks in place of manned kinetic attacks without a significant loss of capability, then that military has an ethical obligation to do so.

As already noted, there will be less risk associated with the remote deployment of cyber weapons than with manned kinetic operations. If weapons are used suffi-

---

[4] And, further, presume that they do not have a scarcity of resources problem such that they must reserve some particular weapons for future missions, etc.

[5] There are, of course, important parallels here to cases of individual self-defense, where proportionality and necessity rule. In fact, many revisionist accounts of just war theory currently on the rise today contend precisely that the moral rules of warfare should track more closely with the moral reality of individual self-defense and, as such, should impose a much stronger "necessity" clause on any given just military action (see, Rodin 2005, for example). Such a discussion, however, is far outside the scope of this paper.

[6] Again, see Steinhoff (2013) for a contrasting view. Presumably Steinhoff's objections to PUR would carry over to PURH.

ciently remotely, there will be little or no risk to the life of military personnel using the weapons. This is the same condition we see obtain with UAVs. Indeed, UAV bomb strikes already presently combine cyber operations (controlling the UAV) with physical operations (dropping the bomb).

In addition, cyber weapons can be less harmful than their physical counterparts. They are generally not lethal and often do not cause any permanent damage to physical infrastructure. Cyber-attacks may even be less damaging than electronic warfare strikes that "fry" electronics such as electro-magnetic pulse (EMP) weapons. A cyber-attack that takes out some service such as telecommunications, power, sensors, or alarms need not cause any permanent damage nor harm anyone. If there is no physical damage, targets are more readily restored to their original state after hostilities have ended. Restoration is faster and costs less. It may just be a matter of restoring bits from backup files, though systems may also have to be patched and security enhanced to avoid future attacks. This can be important for stability and reconstruction operations following war and would lend itself favorably to considerations of *jus post bellum*.[7] Dipert (2010) has argued along similar lines that cyber weapons could even be designed in such a way so that the damage they do is easily reversible, similar to creating an antidote to a real-world contagion. Rowe (2010b) also argues for reversibility and offers four different techniques that could potentially be used to achieve it. Whereas rebuilding critical infrastructure such as telecommunications and electricity power grids can take weeks, months, or even years after being destroyed by *physical* weapons, that same infrastructure could be back up and running within hours or days after a cyber-attack.

As with OP, the operating principle CyberOP presumes that cyber-attacks would be used to take an action that is otherwise deemed just according to the principles of *jus ad bellum* and *jus in bello,* and that using the cyber weapons instead of kinetic ones would not result in any significant loss of capability. The phrase "without a significant loss of capability" is crucial to the formulation of CyberOP, implying that the cyber operations could be sufficiently controlled in such a way that neither risk nor harm would increase in the accomplishment of a given objective. If the deployment of cyber weapons would incur either greater risks to just military forces or cause more harm to adversary forces or non-combatants, then those weapons would thereby be considered less capable than their physical weapon counterparts. In that case, their use would not be mandatory under CyberOP; indeed, their use would likely be impermissible. It's possible they might yet be considered a better alternative on other moral grounds outside the avoidance of unnecessary risk and harm, but further arguments would have to weigh the tradeoffs involved.

Our claims here, of course, do not rest on mere abstract possibilities and speculation about future kinds of weapons. There are cyber weapons which already exist that could potentially fit the demands of CyberOP. Rattray and Healey (2010) give several examples where cyber operations might be used to support special operations or traditional, kinetic military operations. These include using cyber operations to take out adversary alarms or inject false alarms, or to disrupt telecommuni-

[7] See Orend (2000) for a comprehensive case for developing principles of *jus post bellum*.

cations or command and control networks. Conducting these operations remotely using cyber weapons would be less risky to a just force than sending in military personnel to accomplish the same objectives, and likely less damaging to their targets, making them excellent candidates for application of CyberOp.

The principle of CyberOP, however, would not justify many of the cyber warfare scenarios that have been postulated such as, for example, the one in Clarke and Knake (2010, pp. 64–68) that leads to a nationwide power blackout, airline and subway crashes, pipeline explosions, refinery fires, lethal clouds of chlorine gas, network outages, and more, resulting in thousands of civilian deaths. Kinetic strikes that did all this would almost certainly violate the LOAC, ruling out their cyber equivalents. Even an attack that just took down the Internet would likely violate LOAC, as it would necessitate attacking civilian infrastructure around the world, including infrastructure in neutral countries.[8] If one presumes a blanket principle of non-combatant immunity, it is hard to imagine circumstances where such a strike would be considered just under traditional just war theory.[9] However, in those cases where limited kinetic attacks against the Internet might be justified, say to take out a small number of routers in some country for a limited period of time, surely a cyber-attack that temporarily shut down those same routers would be morally preferred over a kinetic strike that physically ruined the routers and killed the persons operating them.

CyberOP would not even justify many lesser operations that have actually taken place, such as the distributed denial of service (DDoS) attacks against Estonia in 2007 that disrupted access to Estonian websites in protest of the relocation of a Soviet-era war memorial (Tikk et al. 2010; Clarke and Knake 2010, pp. 13–16). One reason is that the nature of the dispute did not justify any military action against Estonia to begin with (Lucas 2011b). Since using kinetic weapons against the target websites would not be permitted, using cyber weapons against those targets could not be justified by CyberOP.

Although cyber-attacks need not cause death or physical destruction, attacks that do are not necessarily ruled out by CyberOP. If the harm is no greater than that

[8] For a good discussion on cyber warfare attacking non-combatants and the resulting problems, see Lucas (2011b). Lucas' work in that piece is highly compatible with the claims we make in this paper, although we disagree with Lucas over whether some specific instances of cyber-attacks would be permissible. Lucas writes, "… an act of cyber warfare is permissible if it aims primarily at harming military (rather than civilian) infrastructure, degrades an adversary's ability to undertake highly destructive offensive kinetic operations, harms no civilians and/or destroys little or no civilian infrastructure in the process." And, of course, we further differ from Lucas in contending that in some such instances cyber-attacks would not be merely permissible, but obligatory to use in place of similar kinetic attacks.

[9] Several revisionist just war theorists have recently challenged a blanket principle of non-combatant immunity and have argued that some noncombatants could be liable to harm in war. Jeff McMahan (2009) does this most prominently, but others such as Helen Frowe (2011) have also advanced a rejection of total non-combatant immunity. Note that even on these revisionists accounts, however, an attack against the entire Internet would still fall outside of the bounds of just war practices because there would be very little if any discrimination possible amongst non-combatants. Again see Lucas (2011b) for a discussion on the possibilities for discriminate cyberwarfare.

caused by the just use of kinetic weapons, then the cyber-attack is still preferred, indeed morally obligatory, if it is less risky. Moreover, a cyber-attack that causes equipment to self-destruct, as in the case of Stuxnet, may still be less life-threatening than physical strikes.

Stuxnet is an interesting and important case. On the one hand, the cyber operation enabled the destruction of centrifuges at Iran's Natanz nuclear enrichment facility (Broad et al. 2011) without risking the lives of those who did it nor the operators at Natanz. In that sense, it was less risky and less harmful than, say, dropping bombs on Natanz. On the other hand, Stuxnet caused considerable collateral damage that a bomb strike would not have caused. In particular, tens of thousands of other systems got hit with the worm (Falliere et al. 2011). In that regard, Stuxnet was less capable than a kinetic strike, and so not morally obligatory under CyberOP. This does not mean that Stuxnet was immoral or not preferred over a kinetic strike, only that it does not lend itself to direct application of CyberOP. The morality of Stuxnet is more complicated and we return to it below.

6.3 Objections to Cyber Warfare

Although we believe there is a strong case for conducting cyber-attacks in limited circumstances where they have a moral advantage over kinetic attacks, some have argued that militaries should not conduct cyber-attacks at all. If that is true, then, of course, CyberOP would be a vacuous principle, at best. The following reviews some of these objections. We find that each objection, while important, fails to overcome the strong normative force of PURH and the resulting CyberOP.

6.3.1 Objection 1: Cyberspace Should Not be Militarized

Some argue that military operations should not be conducted in cyberspace, as doing so makes cyberspace less attractive and usable to others, turning it into a perpetual battleground. At any given time, militaries might conduct operations that impair normal activity and harm legitimate use. Everyone who uses the internet would potentially become a target or unwittingly caught in cyber-crossfire.

Our response to this objection is that cyberspace is *already* under constant attack by criminals, protestors, patriotic hackers, cyber jihadists, spies, anarchist groups, and others who pay no heed to legal or ethical constraints on their behavior. By contrast, the militaries of states are or should be concerned with these things, and should conduct themselves under the principles of *jus ad bellum* and *jus in bello*. The cyber military operations we are advocating for in this paper would be so conducted. They are unlikely to even be noticed by most users. Their greatest impact would be felt by legitimate military targets, with less collateral damage than from kinetic strikes. If not and the use of cyber weapons in question was conducted contrary to just war principles, then they would fail to be justified under PUHR to begin with since they would not be a proper case of a good goal G pursued in a manner that does not violate the demands of justice.

A related argument is that military use of cyber weapons runs counter to efforts to make cyberspace more secure and usable. This is because militaries classify their cyber weapons and keep them secret. They will not make their weapons public or report the vulnerabilities they exploit, as doing so would lead to the flaws being repaired, thereby rendering the weapons useless (Rowe 2010a). The net effect of this secrecy is that cyberspace will have unnecessary vulnerabilities that can be exploited not just by militaries but by anyone else discovering them.

Although cyber security is critically important, we do not agree that it is jeopardized by military classification or use of cyber weapons. One reason is that the effects of publicly disclosing vulnerabilities and attack tools are not all positive. While disclosure is likely to enhance security in the long run by removing vulnerabilities, it can also lead to the proliferation of cyber weapons as well as an increase of attacks, as tool developers build on each other's work to create new cyber weapons and criminals take advantage of the lag between disclosure and remediation to launch attacks. Another reason is that if militaries have no vested interest in offensive cyber weapons, they will not allocate resources to cyber weapons development, and hence, will have little information to contribute in the area of cyber vulnerabilities. Moreover, vulnerabilities they do find may be classified anyway in order to protect military systems.

6.3.2 Objection 2: The Deployment of Cyber Weapons will Lead to Their Spread and Use

The deployment of cyber weapons typically has the side effect of making those weapons available to the target and possibly third parties. This is because some or even all of the weapon's code may be present on devices hit by the weapons. For example, when a worm spreads to some computer, its code will be present on the infected computer, making it available to the computer's owner and possibly third parties such as anti-virus companies. As another example, computers that have been compromised and placed on botnets often download additional code in response to instructions from their botnet's command and control facility. All of this code is then available to those with access to the machines.

Although much of the code left behind from cyber weapons will be in the form of executable binaries that are not readily re-purposed, executable code can be reverse-engineered or decompiled into source code, making it more readily available for analysis, reuse, and integration into other code. As a result, the confiscated cyber weapon might be reused or used as a building block for new cyber weapons, perhaps ones that are even more damaging than the original. The new weapons then might be fired back at the source of the original weapon or used to attack other targets. Even if the original tool was used justly, its reuse and offspring might be appropriated for unethical purposes. The net effect can be an increase of damaging cyber-attacks. In addition to spreading covertly as in the above examples, the code for worms, viruses, Trojans, botnets, and other forms of cyber weapons spreads through more overt means. Security researchers, at least many outside government agencies, share information and code (including source code) pertaining to cyber vulnerabilities and tools for exploiting these vulnerabilities. They post it on public websites and sell it through legitimate and black cyber markets. The result has been a proliferation of cyber weapons. The security company Symantec reported that they encountered almost 300 million variants of malicious software in 2010 (Symantec 2011).

Because of these proliferation effects, some argue that militaries should not deploy cyber weapons. There is too much danger that the weapons will get into the wrong hands and be used in harmful ways. Their use will just make the problem of cyber defense worse for everyone. There is an intriguing irony to this argument against the military use of cyber weapons in that it is the inverse of the objection described in the previous subsection, where it was argued that the secrecy of these weapons would limit our ability to learn about and repair vulnerabilities in cyberspace in order to make it more secure. This objection makes the opposite point.

We believe that the proliferation of military-grade cyber weapons is a legitimate and larger concern than their secrecy, especially since such weapons may be more sophisticated than many of the weapons used by other actors in cyberspace. Stuxnet's executable binaries are now out in the public domain, where they have been studied and could be used to develop new weapons. However, militaries can minimize the risk of their cyber weapons falling into adversary and third party hands by precisely targeting them.

Another key way this risk could be mitigated would be for weapons developers to program them to self-destruct after completing their objectives. Although it may not be possible to guarantee complete containment and destruction of military cyber weapons, it may be possible to reach an acceptable level of assurance. If not, then the cyber weapons may not meet the threshold for CyberOP to begin with, as they may be significantly less capable than their kinetic counterparts. We believe that more work needs to be done on developing self-destructing cyber weapons to fight against this proliferation worry. If that is done, and a cyber weapon's proliferation can be contained, then the obligation to employ such weapon can still apply under CyberOP.

6.3.3 Objection 3: Cyber-attacks are too Difficult to Control and Use Effectively

Some argue that cyber-attacks cannot be controlled, and thus could lead to unanticipated and unpredictable harms, including collateral damage. They cite such cyber weapons as viruses and worms, which often spread widely and cause considerable disruption in the process (Rowe 2010a). For example, the Slammer worm shut down ATM machines and emergency 911 systems, caused flight delays, and disabled a safety monitoring system at a nuclear power plant. Even Stuxnet, which limited its primary damage to Iran's nuclear facility, infected tens of thousands of other systems in the process of arriving at and delivering its payload.[10]

We do not accept the premise that cyber weapons cannot be controlled. True, *some* weapons, such as worms that attempt to infect as many devices as possible, seem to be out of control, but we do not anticipate the use of such weapons under CyberOP. That is, any weapons—be they cyber or kinetic—that impose *intentionally* indiscriminate damage in this way would not meet even the most meager adherence to *jus in bello* principles. Rather, we anticipate the use of cyber weapons that are tightly controlled and precisely aimed. Such weapons would have the same or greater level of precision and discrimination as kinetic weapons. Any cyber weapon that could not be controlled would likely fail to meet the principle of CyberOP as it would be less capable than a more controllable and discriminate kinetic weapon. Although kinetic weapons can also cause collateral damage, that damage is relatively limited or, at least, more easily predicted and calculated, compared to cyber weapons that can affect systems all over the Internet.

Rowe (2010a) also argues that cyber weapons are too hard to use effectively. They could be unreliable, as new software systems and weapons often are, with the code failing to work as intended, or an attack failing because assumptions about the target were wrong. Further, the effects of cyber-attacks can be difficult to determine or measure. Even victims of cyber-attacks can have trouble assessing their damages. This is a particularly pernicious problem for the just use of cyber weapons since adherence to *jus in bello* principles of both discrimination and proportionality each require at least some degree of damage predictability for a given attack.

Our position is that cyber weapons whose effects cannot be accurately predicted, controlled, and measured would likely fail to satisfy the conditions for their ethical application under CyberOP. Either they would be less capable than their kinetic counterparts, in which case they would fail to meet the requirements for CyberOP, or else they would have no kinetic equivalents, in which case CyberOP would not apply and additional moral reasoning would be needed to determine whether their deployment is morally just. However, if, as Arquilla (1999, p. 393) argues, the targets of a cyber-attack were strictly limited to strategic military targets, such attacks could very well be morally justified and in-line with CyberOP. We believe this kind of discrimination and control with cyber weapons could be attained.

[10] Although in this case much of that infection did little real damage to systems it used on its way to delivering its payload. This is a more complicated question regarding what should constitute damage and how such a calculus should be used in analyzing the moral permissibility of particular attacks. We address this issue separately below in Objection 4.

6.3.4 *Objection 4: Cyber-attacks Involve Using Unwilling Bystanders as Accomplices for Attacks*

This objection is closely related to Objection 3, but poses its own unique difficulties. Rowe (2010a) claims that military use of cyberspace would produce a kind of unnecessary collateral damage, as militaries would compromise civilian computers in order to place them on botnets or spread worms, or to use them as "stepping stones" or "launch pads" for reaching their targets. We agree that such collateral damage generally should be avoided. However, the objection goes beyond such intentional compromises, as the packets deployed in a cyber-attack could flow through routers and along links owned by private companies and residing in neutral countries. This use of presumably unwilling (and usually unwitting) bystanders (routers) might be considered a violation of the principle of neutrality demanded by *jus in bello*.

Part of this concern arises from the network topology of cyberspace compared with the traditional battle-space of land, sea, or air (Dipert 2010, 2012). When a just force launches and delivers a kinetic weapon to an adversary, they often need not directly involve or traverse various third-party sovereign territories in order to deliver the weapon. Yet cyber-weapons are likely to move through routers in third-party countries on their way to their targets, and the paths they take are difficult to control.[11]

Of course, in the kinetic weapon example, a just force *may* indeed on occasion need to move through a third party's airspace, say, on the way to the target. But if a plane or missile or tank or some such traditional weapon did have to travel through another's territory before arriving at the target, international law would require that they seek permission of the traversed state before doing so. One reason for this is that the traversed state would lose any status of neutrality and, therefore, could be legitimately targeted in a counter attack.

At least in principle, the same permission might be sought for using cyberweapons. However, the situation is considerably more complex, as many sovereign states are likely to be involved in the movement of packets, and packets can flow along different routes and through different countries depending on traffic loads and other dynamic properties of cyberspace. On the other hand, the argument can be made that the movement of packets through third party states does not violate the principle of neutrality. Rather, the situation is the same as for general telecommunications, where belligerents do not need permission to make international phone calls that pass through third party telecommunications switches and links, and the countries providing those switches and links are immune from attack as long as their services are provided impartially to all sides (DoD 1999). Since the routers and links of cyberspace essentially implement the same basic communications relay service, they too should be immune from attack as long as they move the packets of all parties without favoring any side. However, permission would still apply if

[11] Though this need not always be the case. It is possible for a cyber-weapon to directly attack an adversary system without any mediating system whatsoever. But this will be rare.

an attacking state wanted to use the servers of another to launch a cyber-attack or to host files that support the attack.

As we've already made clear, we agree that cyber weapons that produce unnecessary collateral damage should not be used. Indeed, if the kinetic weapons they replace can produce the desired effects but with less collateral damage, then the cyber weapons would be considered less capable and thus non-obligatory under CyberOP. The real difficulty raised by this objection, then, is that such calculations of differing kinds of collateral damage will have to measure not mere quantity of those affected by a just cyber-attack, but the *kind* of harm delivered. It is quite possible that a just cyber-attack that affects a relatively large number of noncombatants as unintended collateral damage, but does so only very minimally (say by very temporarily slightly disrupting their internet access or placing a negligible amount of passive code on their system), could be morally preferable to a just kinetic attack that affected a much smaller number of noncombatants as unintended collateral damage but did so to a much greater degree of harm. If two such approaches were the *only* options available to a just force, it seems at least reasonable that the cyber-attack could be considered the option which produces the least amount of harm, if "least" is meant in terms of severity and not extent. Thereby such a cyber-attack could possibly be justified under PURH and consistent with CyberOP. Such decisions would be difficult, but we do not see any in-principle reason why cyber-attacks should be weighed differently than would two alternative kinetic attacks in similar decisions over weapon choice.

Lucas (2011b) has argued that Stuxnet was justified because of the way its primary damage delivered by the worm was highly discriminate (focusing only on the Iranian centrifuges it was designed to damage) and because an alternative kinetic strike would have done far more physical and, likely, *lethal* harm. He does not consider the other systems infected by Stuxnet to have been harmed. In our view, the non-Iranian systems infected with the Stuxnet worm as part of its delivery should properly be considered collateral damage simply because the owners of those systems could make legitimate complaint that they did not want to have a worm on their system and did not want to have to expend resources removing it and assessing possible damages, all of which can be difficult and time consuming. Further, those infected with Stuxnet on its way to delivery could view it as a violation of their sovereign autonomy over their system in that they unwittingly played the part of accomplice to the attack.

Yet, even if our view regarding the collateral damage of Stuxnet is right, Lucas may still be right that in this case this *kind* of collateral damage is morally preferable to the kind of collateral damage that would have been likely incurred in a kinetic attack on the Iranian nuclear facilities. It that is true, and if a strike designed to impede Iranian nuclear capabilities was otherwise deemed just, then it is possible that Stuxnet could fit the parameters of CyperOP since it would cause less collateral damage than a comparable kinetic strike.

Again, this conclusion would here be taking "less" collateral damage to mean less severe even if not less extensive. Whether that is the morally correct conclusion for how to best weigh different kinds of collateral damage is a matter for another paper and one for which we remain neutral for the purposes of this paper. We do not here address the difficult and complex ways these different kinds of collateral harms should be weighed against one another when alternative means of accomplishing a just attack are available. But, again, we see no in-principle reason why more widely diffused, but less severe, collateral damage could not be morally preferred over more severe damage inflicted to a smaller set of noncombatants.

Note as well, of course, that in this discussion here of Stuxnet we are not arguing that an attack against the nuclear facilities of Iran was necessarily justified to begin with—kinetically or through cyber-weapons—and do not mean to argue for such a conclusion in this paper. Rather, Stuxnet is a good case to examine in light of CyberOP on the *stipulation* that an attack on Iranian centrifuges was an otherwise just attack. Whether that stipulation is *actually* valid is a debate for another paper.

6.3.5 Objection 5: Cyber-attacks could Provoke Unanticipated Responses and Escalate Conflicts

It is impossible to predict with certainty how an attacked state might respond to a cyber-attack. Indeed, some countries, including the United States, have said that they would consider all options, including using kinetic weapons against the attacking state. A state might even respond with the nuclear option. Clearly, such actions could escalate a conflict.

Because a cyber-attack might provoke a retaliatory cyber or physical strike far in excess of the original attack, some have argued that cyber-attacks should not be used at all. However, any action can potentially provoke an unanticipated, harmful, and disproportionate response. Even the relatively non-aggressive plan to relocate a war memorial in Estonia provoked not only the DDoS attacks against Estonian websites, but also riots in the streets (Tikk et al. 2010).

This is not to say that because all responses cannot be anticipated, they should be ignored when conducting cyber-attacks. Rather, it is to argue that the possible effects of all types of actions should be considered. There is no reason to single out cyber-attacks as being more likely than physical attacks to lead to severe retaliatory strikes with conflict escalation. Indeed, because cyber-attacks are often difficult to attribute and hence deniable by their perpetrators, they might be less likely to provoke a retaliatory strike. The targeted state may not be sure who to retaliate against, and so proceed on the side of caution rather than risking an unprovoked counter strike against innocent parties. Even if the target correctly identifies the origin of the cyber-attack, it might retaliate with an in-kind cyber-attack, which may be less harmful than had it chosen a kinetic strike. The general non-lethal and less destructive nature of cyber-attacks gives further reason to predict that nations will respond with less damaging counter attacks, if they respond at all. Indeed, as has been noted, the very nature of cyber-attacks often makes it possible for a given target to recover from an attack quickly and with little or no permanent damage. This would allow a target nation to "save face" and either deny the attack or downplay the damage it caused.[12] Such scenarios, it seems to us, would be less likely to result in conflict escalation than comparable kinetic strikes would (Lucas 2011b; Owens et al. 2009). If not, they would at least allow for an "escape valve" to avoid direct kinetic hostilities in ways that a comparable kinetic strike would not.

6.3.6 Objection 6: Cyber Weapons make Warfare too Easy

Some argue that cyber weapons makes warfare too easy. Whereas states may be reluctant to conduct physical strikes, they may be less reluctant to conduct remote attacks in cyberspace. The result could be the launching of more wars, including wars that use both cyber and physical weapons, than would otherwise happen. The fear here is not merely more wars but, most likely, when more wars are launched, there will be more unjust wars. This is the familiar "threshold" problem for all new advances in military technology (Strawser 2010; Lucas 2011a). The worry is that by making war too easy, cyber weapons will entice states to undertake war in violation of the restrictions normally imposed by the *jus ad bellum* principle of last resort.

Our response is that the principle of CyberOP protects against this and blocks the objection. It assumes that cyber weapons are used in a context where a comparable physical action is already deemed just under the traditional just war convention and LOAC. Thus, the cyber operation should not be construed as in illegal act of war violating the principle of *jus ad bellum* or *jus en bello*. It should not be seen as an illegitimate act of force or aggression, and should not lead to a full-scale, unrestricted war. Further, as was argued in Strawser (2010, pp. 358–360), this "threshold" problem is a difficulty for every new advance in military technology; it is not unique to cyber weapons just as it is not unique to UAVs. And if a given technology is used justly in a present case and is the morally obligated weapon choice due to considerations of unnecessary risk or harm, then mere speculation about its future misuse should not trump the present normative obligation to so use it.

Notice that there is, in fact, potential for moral gain here vis-à-vis the inverse of this objection. The moral worry raised by this objection is that the ease with which a nation-state can use cyber weapons lowers the threshold to resort to war and could thereby result in more wars, which presumably means more unjust wars. But the inverse could also be true: that just causes that *should* be persecuted but are not, could be carried out by nation-states willing to use cyber-weapons who would not be willing to use kinetic weapons (even if they should).[13] In our view, this cannot stand alone as a positive argument for the development and employment of cyber-weapons, because it is equally possible that they could be used for nefarious ends.

---

[12] This is exactly what played out with the Stuxnet attack on Iran.

[13] Savulescu and Beauchamp (2013) argue for a similar moral gain that could be possibly had with regard to the increasing use of UAVs.

But if cyber-weapons are used in line with the strictures of CyberOP, such use could result in precisely this kind of normative gain because states may be more willing to use cyber weapons due to the advantages they provide with regard to force protection through the avoidance of unnecessary risk.

In addition, we expect states following the obligations of CyberOP to be cautious about using cyber weapons, because of the uncertainty about how a target might respond. In that regard, cyber weapons serve as a deterrent against state use, not because their use would be so devastating (as with nuclear weapons), but because of the uncertainty that the target might respond in a manner that is devastating (say by using nuclear weapons). Thus, we would expect militaries to use cyber weapons more for surgical strikes conducted in the context of just wars, and for just covert operations having limited effects such as, for example, disabling local communications, power, or alarms long enough for a hostage rescue mission to complete successfully.

6.4 Conclusions

We do not claim that all cyber weapons are ethical in principle. Indeed, many are not by their very design. Rather, our claim is that in limited circumstances and granting certain assumptions, cyber weapons are morally preferred over their kinetic counterparts resulting in an obligation to use them in line with traditional just war theory principles. In particular, they are a better option when they can be deployed for a purpose already deemed just under LOAC and without any significant loss of capability. This moral preference arises out of the simple moral obligations imposed by the PURH. That is, cyber-attacks can be less risky and harmful than kinetic strikes, and can thereby impose a duty for militaries to use cyber weapons in place of their kinetic counterparts.

We leave open the ethical questions surrounding the use of cyber weapons that do not have apparent kinetic counterparts and hence are not covered by CyberOP. An example would be a cyber weapon that alters data on an adversary system so as to present false information to the adversary. Stuxnet did this. In addition to altering the code driving the centrifuges so as to physically damage them, it altered the data displayed to the operators so as to hide the effects of the attack. The ethics of this and other operations not covered by CyberOP requires additional moral reflection, including consideration of basic LOAC principles, in order to determine whether the operations are just.

We also leave open the ethics of using cyber weapons that have some of the capabilities found in kinetic weapons, but not their full capabilities. It may be that such cyber weapons are still ethically superior because of other capabilities that are not present in the kinetic weapons. The moral permissibility of such weapons would depend crucially on their ability or lack thereof to be used in a just manner as part of an overall just attack.

Although we have focused on the conduct of cyber-attacks rather than cyber exploitation (espionage), the same general reasoning might apply to exploitations. In particular, when foreign intelligence can be collected through a cyber-operation as opposed to one that requires physical presence in foreign territory or the turning of a foreign insider (e.g., to leak classified documents), then the cyber operation might be preferred and perhaps even morally obligatory on the grounds that it would be less risky to collectors and less harmful to those collected against. However, we leave a thorough analysis of this for future study.

References

Arquilla, J. 1999. Ethics and information warfare. In *The changing role of information in warfare,* eds. A. Khalilzad, J. White, und A. Marshall, 379–401. Santa Monica: RAND Corporation.

Broad, W. J., Markoff, J., and Sanger, D. E. 2011. Israeli test on worm called crucial in Iran nuclear delay. *The New York Times*, January 15.

Clarke, R. A., and Knake, R. K. 2010. *Cyber war.* New York: Harper Collins.

Denning, D. E. 2008. The ethics of cyber conflict. In *The handbook of information and computer ethics,* eds. K. E. Himma and H. T. Tavani, 407–428. Hoboken: Wiley.

Dipert, R. 2010. The ethics of cyberwarfare. *Journal of Military Ethics* 9 (4): 384–410.

Dipert, R. 2012. Ethical aspects of cyberwar. Paper presented at the meeting on ethical and societal issues in National Security Applications of Emerging Technologies for the National Academy of Sciences Division on Engineering and Physical Sciences, Computer Science and Telecommunications Board, Washington D.C.

DoD. 1999. *An assessment of international legal issues in information operations*. 2nd ed. (November). Arlington: Department of Defense, Office of General Counsel.

Falliere, N., Murchu, L. O., and Chien, E. 2011. W. 32 Stuxnet Dossier, V. 1.4, Symantec Security Response. February.

Frowe, H. 2011. Self-Defence and the principle of non-combatant immunity. *Journal of Moral Philosophy* 8:530–546.

Lucas, G. R. 2011a. Industrial challenges of military robots. *Journal of Military Ethics* 10 (4): 274–295.

Lucas, G. R. 2011b. *Permissible preventive cyberwar: Restricting cyber conflict to justified military targets*. Oxford: Oxford Institute for Ethics, Law, and Armed Conflict.

McMahan, J. 2009. *Killing in war*. Oxford: Oxford University Press.

Orend, B. 2000. Jus Post Bellum. *The Journal of Social Philosophy* 31 (1): 117–137.

Owens, W. A., Dam, K. W., and Lin, H. S., eds. 2009. *Technology, policy, law, and ethics regarding US acquisition and use of cyberattack capabilities*. Washington, DC: The National Academies Press.

Rattray, G., and Healey, J. 2010. Categorizing and understanding offensive cyber capabilities and their use. In *Proceedings of a workshop on deterring cyberattacks,* eds. K. W. Dam and W. A. Owens, 77–97. Washington, DC: The National Academies Press.

Rid, T. 2011. Cyber war will not take place. *Journal of Strategic Studies* V 35, no 1, 5–32, February, DOI:10.1080/01402390.2011.608939.

Rodin, D. 2005. *War and self-defence*. Oxford: Oxford University Press.

Rowe, N. C. 2010a. The ethics of cyberweapons in warfare. *Internationl Journal of Technoethics* 1 (1): 20–31.

Rowe, N. 2010b. Towards reversible cyberattacks. In *Proceedings of the 9th European Conference on Information Warfare and Security,* ed. J. Demergis, 261–267. Reading: Academic Publishing Ltd.

Savulescu, J., and Z. Beauchamp. 2013. Robot angels: The use of UAVs in humanitarian military intervention. In *Killing by remote control: The ethics of an unmanned military,* ed. B. J. Strawser, 106–125. New York: Oxford University Press.

Schmitt, M. N. 1999. Computer network attack and the use of force in international law: Thoughts on a normative framework. *Columbia Journal of Transnational Law* 7:885–937.

Schmitt, M. N. 2010. Cyber operations in international law: The use of force, collective security, self-defense, and armed conflicts. In *Proceedings of a workshop on deterring cyberattacks,* eds. K. W. Dam and W. A. Owens, 151–178. Washington, DC: The National Academies Press.

Steinhoff, U. 2013. Extreme asymmetry and its discontents. In *Killing by remote control: The ethics of an unmanned military,* ed. B. J. Strawser, 179–207. New York: Oxford University Press.

Strawser, B. J. 2010. Moral predators: The duty to employ uninhabited aerial vehicles. *Journal of Military Ethics* 9 (4): 342–368.

Symantec. 2011. Internet security threat report. Trends for 2010, Vol. 16, April.

Tikk, E., Kaska, K., and Vihul, L. 2010. *International cyber incidents: Legal considerations.* Estonia: NATO Cooperative Cyber Defence Centre of Excellence.

Wingfield, T. 2000. *The law of information conflict.* Falls Church: Aegis Research Corporation.

Wingfield, T. 2009. International law and information operations. In *Cyberpower and national security*, eds. F. D. Kramer, S. H. Starr, and L. K. Wentz, 525–542. Washington, DC: NDU Press.

Chapter 7
The Ethics of Cyberattack

Steven P. Lee

Abstract The internet has made it possible to do damage at a distance by the use of networked computers. A deliberate act doing such damage may be referred to as a *cyberattack*. My concern in this essay is the ethics or morality of cyberattack as a part of war. The morality of war or military attacks in general is judged in terms of just war theory, which examines war in its two aspects, the morality of going to war (*jus ad bellum*) and the morality of conduct in war (*jus in bello*). I examine the morality of cyberattacks in each of these areas. My conclusion is that, while the use of cyberattacks is a novel form of conflict in many ways, its ethical dimensions can for the most part be understood in terms of the traditional categories of just war theory. There remains, however, an important aspect of cyberattack that may carry us beyond the limits of traditional just war thinking about war.

The internet has made it possible to do damage at a distance by the use of networked computers. A deliberate act doing such damage may be referred to as a *cyberattack*. In the words of one study: "Cyberattack refers to deliberate actions to alter, disrupt, deceive, degrade, or destroy computer systems or networks or the information and/or programs resident in or transiting these systems or networks" (National Research Council 2009, p. 1). Cyberattacks (also called computer network attacks) are a species of information operations. Cyberattacks may be carried out for a variety of purposes, just as ordinary (non-cyber) attacks may be. For example, there is cybercrime consisting of cyberattacks, as there is ordinary crime consisting of ordinary attacks of various sorts. When a series of ordinary attacks is carried out by a state against the interests of another state, this is sometimes a conventional, shooting war. Cyberattacks too may also be carried out by states against the interests of other states. Randall Dipert notes that cyberattacks may be "coordinated by the central commands of governments (or other political organizations), and [may be] directed at another country's governmental and military information systems, or at its commercial or infrastructure information systems for political purposes" (Dipert 2010,

S. P. Lee (✉)
Hobart and William Smith Colleges, Geneva, USA
e-mail: lee@hws.edu

L. Floridi, M. Taddeo (eds.), *The Ethics of Information Warfare*,
Law, Governance and Technology Series 14, DOI 10.1007/978-3-319-04135-3_7,

p. 385). When a state is, in this way, directing cyberattacks against another state, the result may be a *cyberwar*.[1]

My concern in this essay is the ethics or morality of cyberattack. The morality of war or military attacks in general is judged in terms of just war theory, which is the millennia old intellectual tradition in the West for assessing war in moral terms. Just war theory examines war in two respects, the morality of going to war, which is traditionally referred to as *jus ad bellum*, and the morality of fighting in war, referred to as *jus in bello*. So an ethical examination of cyberattacks should consist in considering cyberattack from each of these perspectives. In the first section, I examine some general issues on the nature of cyberattacks and cyberwar. The second section is devoted to the consideration of cyberattacks from the *ad bellum* and *in bello* perspectives, and the third section raises some further ethical issues raised by cyberconflict. My conclusion will be that while cyberattack is a novel form of conflict in many ways, its ethical dimensions can for the most part be understood in terms of the traditional categories of just war theory. There remains, however, an important aspect of cyberattack that may carry us beyond the limits of traditional just war thinking about war.

7.1 Section I

There is no doubt that cyberattacks can have an *operational role* in conventional war, that is, they can be (and have been) used in conventional warfare, for example, to disrupt the opponent's military communications (National Research Council 2009, p. 2). But some have argued that because cyberattacks do little or no damage in the physical world (as opposed to cyberspace), they are not sufficiently destructive by themselves to initiate or constitute a war. We may express this point by saying that stand-alone cyberattacks are not acts of war and that there can be no such thing as a cyberwar, understood as a war consisting largely or exclusively of cyberattacks.[2] Whether there could be a cyberwar in this sense may seem a merely verbal matter, but the answer to the question has important normative implications, which makes it worth our consideration.

---

[1] I do not have much to say in this paper about the use of cyberattacks by non-state agents, because, as I claim later, the likelihood that such attacks could rise to the level of acts of war is not significant.

[2] As a point of comparison, note that some would, for a very different reason, deny that there could be a nuclear war, understood as a major war consisting largely or exclusively of nuclear attacks. They would argue that "nuclear war" is a misnomer on the grounds that it must be possible for a war to have winner in the traditional sense, which a large-scale nuclear conflict would not have. There could not be a "nuclear war" because nuclear attacks are too destructive, while there could not be a "cyberwar" because cyberattacks are insufficiently destructive.

On a standard definition, war is the use of armed force for political purposes by one state in a large-scale conflict with another state.[3] International law makes the use of armed force a necessary condition for war (Schmitt 1998–1999). The main objection to the idea that there could be cyberwar is that cyberattacks do not conform to this definition. Cyberattacks, it is claimed, do not involve a use of armed force. No force is used in a cyberattack, and computers are not "arms" (Dipert 2010, p. 396). Ordinary war takes place in the physical world involving *kinetics* and physical damage. A cyberattack by itself kills no one; it is a matter of disruption rather than destruction. Note that in the definition of cyberattack in the opening paragraph, the harm that cyberattacks do is to cyber networks themselves and the data they contain, not to anything in the physical world apart from computer hardware. In addition, cyberconflicts take place in "cyberspace," which is different from physical space. In this sense, a cyberattack involves no crossing of borders, which are markers in physical space, and no violation of sovereignty understood as territorial integrity.[4]

One proponent of the argument that cyberconflict is not war is Thomas Rid. He argues that a cyberconflict is not a war because it fails to satisfy the three conditions necessary for war, conditions similar to those in the definition above. Rid argues that a war must be lethal, instrumental, and have a political goal. He argues that the stand-alone episodes of apparently state-sponsored cyberattacks to date have all been examples of subversion, espionage, or sabotage. No act in these categories, he claims, satisfies the three conditions, so none of these episodes has been by itself an act of war (Rid 2011, p. 2). The term "cyberwar," he asserts, involves a metaphorical usage of "war," as in the phrases "war on obesity" or "war on cancer." He suggests that there is a spectrum of activities between crime at the one end and conventional war at the other. State-sponsored cyberattacks with a political motive reside in the middle of this spectrum (Rid 2011, p. 3). Like other examples of subversion, espionage, or sabotage, they are the sorts of acts states may commit against each other outside the context of war. Others have, like Rid, made the claim that cyberattacks, taking place in cyberspace, are nonlethal (Bayles 2001, p. 47). Without the kinetics of regular war, there is little or no physical damage.

Were stand-alone cyberattacks not acts of war, a normative implication would be that they would not be covered under the law of war. While they might still be covered under other aspects of international law, these aspects might be weaker or more controversial in their application. The result might be that states would be substantially free to pursue a broad array of cyberattacks without contravention of their obligations under international law (Schmitt 1998–1999, p. 935; Schmitt 2002, p. 396).

---

[3] For more general purposes, revisions would have to be made in such a definition to account for civil war in its various forms. Later I will address the role of non-state agents in cyberconflict.

[4] When a series of cyberattacks were aimed at Estonia in 2008, NATO refused Estonia's request to invoke the collective self-defense provision of the NATO treaty on the ground that its sovereignty had been violated, stating that "a cyber attack is not a military action" (Lucas MS, p. 9).

But the argument that stand-alone cyberattacks are nonlethal and largely harmless, and so cannot be acts of war, is not sound. The argument depends on one or another of two implausible premises (Rid seems to rely on both of them). The first questionable premise is that we should expect that cyberattacks will do little physical harm because they have to date done little physical harm. The second relies on a cramped understanding of what counts as an effect of a cyberattack. Regarding the first premise, while it is true that cyberattacks have to date not done much physical damage, there is no reasonable expectation that the will continue in the future. The military application of cyber technology has not yet matured. The recent public concern about cyberattacks is due precisely to the reasonable belief that in the future cyberattacks will be able to do a great deal of harm. Indeed, this has already occurred. The Stuxnet computer worm has reportedly done serious physical damage to centrifuges being used by Iran to enrich uranium. Referring to this cyberattack, Michael Hayden, former head of the American CIA said, "Previous cyberattacks had effects limited to other computers.... This is the first attack of a major nature in which a cyberattack was used to effect physical destruction." He concluded: "Somebody crossed the Rubicon." (Quoted in Sanger 2012) The second premise relies on a bogus distinction between direct and indirect effects. The claim that cyberattacks are inherently nonlethal is like a claim that shooting a rifle is nonlethal because all it does is send a projectile through the air. Cyberattacks have the potential to do a great deal of damage (albeit indirect) in the real world, including the loss of human life. As the US government notes: "Critical life-sustaining infrastructures that deliver electricity and water, control air traffic, and support our financial system all depend on networked information systems" (Whitehouse 2011, p. 3). When such systems are deliberately attacked, the damage can be severe. While not all cyberattacks would have lethal effects, many would have lethal effects, and, more importantly, many would be intended to have lethal effects. Joseph Nye notes: "Major states with elaborate technical and human resources could, in principle, create massive disruption as well as physical destruction through cyber attacks on military as well as civilian targets" (Nye 2011, p. 21).

To give an example of one possible future scenario for a series of cyberattacks on the United States, consider the case sketched by authors William Clarke and Robert Knake (Clarke and Knake 2010, pp. 64–68). Fires have erupted at oil refineries across the nation, major gas pipelines have exploded, and toxic clouds of chlorine gas have been released from chemical plants. Air traffic control systems have collapsed, leading to multiple airline crashes, and train routing systems have failed, leading to multiple crashes and derailments. Signal lights have failed, resulting in accidents and massive gridlock in major urban areas. A power blackout covers the entire nation, and natural gas is not flowing, leaving millions in the cold. The economic system is completely frozen due to the elimination of financial data on central computers, and ATMs will not function. The networks of the Department of Defense, both classified and unclassified, have crashed, leaving the military a set of isolated units. Thousands would have died in the space of a few hours, and many more would do so in the days ahead as the effect of food and power shortages take their toll.

Clearly such a deadly cascade of effects from cyberattacks should be counted as an act of war. What is needed to recognize this reality is a focus not only on the means by which an attack achieves its effects, such as whether the deed is done by bombs or by computers, but also on the effects themselves. The effects of an attack play a significant role in determining whether the attack should be treated as an act of war, making just war theory and international law relevant to its assessment. In a study of cyberattacks, the National Research Council noted that the application of the terms force and armed attack "should be judged primarily by the effects of an action rather than its modality" (National Research Council 2009, p. 3). But the means or modality by which the effects are achieved should not be completely ignored. Michael Schmitt suggests the importance of appealing to consequences, but he rejects an exclusive reliance on consequences to determine what counts as an act of war. For example, he points out that economic and political coercion can have many of the negative effects of acts of war, though they are not treated by international law as acts of war (Schmitt 1998–1999, p. 908; National Research Council 2009, p. 257). For example, the economic sanctions on Iraq in the 1990s, on one estimate, led to the deaths of 239,000 children under five (Powel 1998). But there was no war in a legal sense waged against Iraq during most of the 1990s. The lethality of economic sanctions is distinct from the lethality of armed force, independent of the magnitude of the consequences. The question is on which side of this distinction the lethal effects of cyberattacks belong. Are they more like the effects of economic sanctions or more like the effects of armed force?

Clearly not all cyberattacks would count as acts of war. Cyberattacks cover a wide range of types and degrees of intrusion, and many of them are not even potentially lethal. In terms of types of attacks, some are passive and some are active. The passive intrusions may be intended simply to collect information (as in the case of an espionage attack, mentioned by Thomas Rid), while the active attacks are intended to affect or damage a computer system (and thereby often do damage in the physical world). Active intrusions can range from seeking to gain access in order to control a computer system, to implanting computer viruses or worms to destroy or corrupt data, to planting a "logic bomb" that is intended to lie in wait in a system ready to "explode" and do damage upon an internal or an external signal[5] (Schmitt 2002, p. 367). But, more to the point, the active intrusions can be intended or can achieve different degrees of physical damage. Schmitt claims that cyber attacks may or may not be acts of war, "depending on their nature or likely consequences" (Schmitt 2002, p. 375).

In order to distinguish cyberattacks that are acts of war from those that are not without appealing exclusively to consequences, Michael Schmitt seeks to determine the proper extension for the term "armed force." He proposes a "consequence-based interpretation" of the term. He claims that "the reference to armed forces is more

[5] Schmitt, "Wired Warfare," p. 367. The distinction between active and passive intrusions may be represented by the contrast between the Stuxnet worm (June 2010), which sought to damage nuclear centrifuges in Iran and the Flame virus (May 2012), apparently meant simply to collect information.

logically understood as a form of prescriptive shorthand for activity of a particular nature and intensity" (Schmitt 2002, p. 371, 396). The prescriptive shorthand implicitly takes into account not only the human suffering caused by an attack, but also the severity, immediacy, directness, and invasiveness of that harm. The use of armed force tends to have these characteristics to a high degree, while the use of economic and political sanctions does not, despite the fact that both may cause a great deal of human suffering. "Thus, the consequences of armed coercion are presumptively impermissible, whereas those of other coercive acts are not (as a very generalized rule)"[6] (Schmitt 1998–1999, pp. 914–915). This is the basis, in his view, of the distinction between harm imposed on one state by another that should count as an act of war from such harm that should not. Because the harmful effects of cyberattacks may be devastating, immediate, direct, and invasive, as in the scenario from Clarke and Knake (2010) presented above, it follows that cyberattacks can sometimes be acts of war. While Thomas Rid is correct to claim than most cyberattacks (including most of those that have occurred to date) are not acts of war, he is wrong to conclude that cyberattacks cannot be acts of war and that there cannot be a cyberwar. Standalong cyberattacks, specifically those that fall into his category of sabotage, may, if severe enough, be acts of war.

So a cyberattack can be an act of war, and a war (a cyberwar) may be composed exclusively of cyberattacks. One interesting question that Schmitt's analysis seems to leave open is whether a stand-alone *operational* cyberattack could be an act of war. The Clarke scenario is an example of a *strategic* cyberattack, that is, one directed against the economic and social foundations of a society, but an operational cyberattack would be directed against military computers, attacking military command and control.[7] It seems that a large-scale operational cyberattack should count as an act of war. In terms of conventional war, an operational attack is a paradigm act of war, whereas a strategic attack is, in a sense, aberrational. If a strategic cyberattack counts as an act of war, so should an operational cyberattack. But, Schmitt's analysis seems to preclude an operational attack, at least one involving little collateral damage, being an act of war, the reason being that he places human suffering at the center of his case that a cyberattack may be an act of war. Consider this apparent paradox: cyber technology[8] promises the possibility of a major operational attack being achieved with much less human suffering than a conventional operational attack. The goal of an operational attack is disruption of the opponent's military, which a cyberattack might achieve by damage to the relevant computer systems and little harm to humans,[9] while a conventional operational attack, even with highly

[6] Given the potential severity of economic sanctions, as the Iraq sanctions indicate, the way might be open to challenge this categorization by positing that economic sanctions can also sometimes be acts of war.

[7] On the idea of strategic and operational cyberattacks, see (Schmitt 2002, p. 366; Arquilla 1999, p 389).

[8] I use the term "cyber technology," henceforth, to refer to the use of such technology for military purposes.

[9] For a discussion of this sort of cyberattack, see (Bayles 2001, p. 50).

accurate munitions, would blow things up and kill people. Just war theory is concerned to keep human suffering at a low level. So, the better an operational cyberattack would look in terms of just war theory, the less Schmitt's analysis would view it as an act of war. The result would be, at the extreme, that no cyberattack Schmitt would count as an act of war could be just, a result we should not accept. So, whatever the implications of Schmitt's analysis, I will regard a major operational cyberattack, not only a major strategic attack, as an act of war. (In what follows, I will use the term "cyberattack," unless otherwise qualified, to refer to an attack with consequences sufficiently serious to count as an act of war[10]).

Assume that one state launches a major cyberattack, whether strategic or operational, against another state. Where might things go from there? The state attacked might respond with a cyberattack, and if the war continued through a series of cyber exchanges, the result would be a cyberwar. But it seems more likely that such a war would become a conventional war, that the exchanges would move at some point from the cyberspace of the initial attack into physical space. The war would involve real bombs as well as logic bombs. This likelihood is important in understanding the moral assessment of cyberattacks, to which the next section is devoted.

7.2 Section II

Cyberattacks that are acts of war, as well as cyberwar more generally, like their conventional counterparts, are subject to normative regulation. There are rules of war, and some of these rules are moral or ethical, specifically, the rules of just war theory. This section discusses the applicability of the rules of just war theory to cyberattacks. The main question is whether the moral rules that apply to war in general are adequate or relevant to the phenomenon of cyberconflict. Does the moral theory traditionally applied to war address the novel moral issues raised by cyber technology? Does just war theory provide practical guidance to the cyberwarrior? Through its long history, just war theory has had to weather many social, political, and technological changes in the nature of war, changes which threatened to make the theory irrelevant or inapplicable in practice. But the theory has endured these changes largely intact, remaining relevant through many revolutions in military affairs. Can the same be said in regard to the technological changes that have created the possibility of cyberattack?

First, consider cyberattacks from the *ad bellum* perspective *Jus ad bellum* is composed of the moral rules concerning the initiation of war. Can these rules make sense of our moral choices of initiating a war through cyberattacks? The *ad bellum*

[10] As a terminological point, it should be noted that *any* cyberattack in the context of a conventional war could be referred to as an act of war. The claim in this paper that only cyberattacks causing sufficient physical damage would be acts of war refers to *stand-alone* cyberattacks, cyberattacks outside the context of a general war. This latter includes cyberattacks that initiate a war, a war which might then either continue as a cyberwar or become a conventional war, for example, if there was a conventional retaliation to the initial cyberattack.

rules are usually represented by a series of six criteria that must be satisfied before it is justified to go to war. A state is justified in initiating a war only if the war (1) has a *just cause*, (2) is declared by a *legitimate authority*, (3) is begun with a *rightful intention*, (4) shows *proportionality* between means and ends, (5) has a *reasonable chance of success*, and (6) is a *last resort*. Can these criteria represent an adequate standard by which to judge cyberattacks and cyberwar, as they do in regard to conventional war?

Some of the criteria are clearly problematic in the cyber context. Consider (1) just cause, often considered the most important of the *ad bellum* criteria. The paradigm just cause for a state's going to war is the opponent's aggression. A state is justified in using armed force when this is in defense against an act of aggression. The initial act of war may be a conventional attack or a cyberattack. But two problems arise in determining whether a state has a just cause when the initial attack to which it responds is a cyberattack. First, there is a *threshold problem*, and second there is the *attribution problem*. The threshold problem is that the initial attack must be an act of war, as opposed to a lesser act of force. If an attack using cyber technology falls short of an act of war, the state under attack has no just cause to go to war. The problem is exacerbated in the case of cyberattack due to the difficulties discussed earlier about distinguishing cyberattacks that are acts of war from those that are not. But this is a problem with conventional attacks as well; for example, a few shots fired across a border would not normally be an act of war and would not be a just cause for going to war. So the threshold difficulty is not a problem new to cyber technology.

The attribution problem is the difficulty of determining the source of an attack (Dipert 2010, p. 385, 401). Given the nature of cyber technology, it is often difficult to determine where at attack has come from and to justify this to the world, especially when the attacker seeks to hide its identity (Rowe 2010). This raises epistemological questions about the degree of certainty a state must have about who its attacker is before it has a just cause to respond with force (Dipert 2010, p. 393). In the case of conventional attacks, by contrast, the source of a major attack is usually obvious. John Arquilla claims, however, that the attribution problem in the case of cyberattack "is indeed difficult but is not insurmountable." He notes that attribution is sometimes a problem in the case of conventional attacks as well, for example in the historical case of the source of "phantom" submarine attacks on merchant ships bringing supplies to the Loyalists in the Spanish Civil War. He shows how this case was effectively dealt with (Arquilla 1999, p. 396). It may arise as well in the case of conventional terrorist attacks by state-sponsored agents (or in an earlier era, by pirates). Arquilla argues that one can make inferences based on the purposive nature of acts of war, along with other detection techniques, allowing a state usually to discover with a sufficient degree of certainty the source of anonymous attacks. We will revisit the attribution issue in the next section.

But there is another dimension to the attribution problem, namely, the disruptive cyber activity of independent individuals commonly referred to as hackers. Arquilla speaks of the way "in which the information revolution empowers small groups and individuals to wage information warfare" (Arquilla 1999, p. 394). The means of cyberattack are inexpensive and widely distributed, so that anyone with the proper skills can engage in disruptive or damaging activities over the internet. Arquilla himself raises the problem of hackers not in connection with the attribution problem, but instead in connection with the criterion of (2) legitimate authority. This criterion requires that anyone engaging in war have the authority within a large political organization such as a state to do so. Hackers obviously do not have legitimate authority in regards to acts of war. But the existence of hackers does not pose a problem for legitimate authority. *Ad bellum* rules require that those making war be legitimate authorities. But hackers are not making war, since war is a conflict between large political organizations with lines of authority. That hackers may cause a lot of damage in not a problem for the rules of war. It is rather a problem of law enforcement (as is the problem of pirates when they are not state-sponsored).

But hackers may represent a dimension of the attribution problem for the criterion of just cause. If a state finds itself under cyberattack, and if the attack might have come from hackers rather than from a state, the problem of attributing the attack to a particular state for the sake of establishing a just cause for a military response is obviously more difficult. But how much does the hacker phenomenon exacerbate the attribution problem? There are two views on this. One is connected with the common perception that cyberattacks are a weapon of the weak, an equalizer between the weak and the strong, whether the weak happen to be a state, a small independent group, or an individual hacker.[11] On this view, the relatively powerless, including hackers, can through cyberattack do outsized damage to powerful states. Joseph Nye endorses this view. He offers a supporting quotation from a US military official ("Sooner or later, terror groups will achieve cyber-sophistication.") and cites another who argues that "while states have the greatest capabilities, nonstate actors are more likely to initiate a catastrophic attack" (Nye 2011, pp. 21–22).

Another view is that the production of effective cyber weapons is an "expensive, skilled, labor-intensive [and] state-centric enterprise" (Lucas Cyberwar, p. 18). Hackers can be disruptive, shutting down websites and such, but cannot do the high level of physical damage that would be equivalent to an act of war, whether strategic or operational. As evidence for this perspective, one could cite the Stuxnet worm, designed to interfere with the centrifuges Iran was using to refine uranium. The widespread view is that the complexity and sophistication of Stuxnet required that it be produced with the resources of an advanced state[12] (Lucas Cyberwar, pp. 14–16). Even so, might the hackers catch up over time, as Nye suggests? Perhaps, but they would be aiming at a moving target. Strong states will be developing their defensive as well as their offensive capabilities, and any increase in offensive capability by the hackers may be more than compensated for by their targets' increase in defensive capability. If this second view is correct, then the activities of non-state hackers (or, to a lesser extent, weak states) do not add greatly to the attribution problem. A sophisticated cyberattack will reveal the hand of a powerful state. States may seek to

[11] This view is expressed, for example, in (Schmitt 1998–1999, p. 897).

[12] This view was confirmed by a news article documenting how Stuxnet was a project of the United States and Israel (Sanger 2012).

use apparently independent agents, so called patriotic hackers, to mask their identity (Arquilla 1999, p. 387). But a (relatively) sophisticated attack could justifiability be attributed to a state, whether the attack came directly from the state or from apparently independent agents the state is using to cloak its involvement. If an attack is (relatively) unsophisticated, it may be assumed to come from a non-state source, in which case seeking out the hackers to hold responsible would be a matter of criminal law (Bayles 2001, p. 55).

What about the other *ad bellum* criteria? Criterion (3) rightful intention requires that a war be initiated with the intention to address the just cause for the war, and this seems to apply to cyberattacks in a straightforward way. In addition, criterion (5) reasonable chance of success, which requires that a war not be a hopeless cause, also seems to apply unproblematically to wars initiated by cyberattack. Criterion (4) proportionality requires that a war be reasonably expected to produce more good and harm. Its application to cyberattack may not be so clear. On the one hand, proportionality seems potentially more easily satisfied by a cyber war, given that cyberattacks are in general easier to carry out with a minimal loss of life than conventional attacks. On the other hand, there is a factor that militates against this. In general, cyberweapons cannot be tested, because to test them may be to reveal to the opponent how it needs to adjust its systems to defend against that mode of attack, for example, what antivirus patch it needs to develop. For this reason and others, it may be unusually difficult to predict how effective a cyberattack would be. As a result, states may tend to err on the side of a larger response, which would make proportionality more difficult to satisfy (Rowe 2010).

The most serious problem posed by cyberattack to the *ad bellum* rules may be arise in the case of the criterion (6) last resort. This criterion requires that a state go to war only if it has no reasonable peaceful alternative. This criterion is crucial to the success of just war theory in limiting the occurrences of war. There may often be cases where a state has a just cause to go to war (and where other criteria are satisfied as well), but where there are peaceful alternatives that may resolve the conflict. The purpose of the last resort criterion is to insure that war is not resorted to in such cases, at least until peaceful alternatives have been shown to fail. Arquilla argues that this criterion is one of the respects in which cyberattack technology plays havoc with the traditional morality of war, leaving "just war theory in tatters" (Arquilla 1999, p. 394). The main way in which cyber technology undermines the applicability of the last resort criterion is in the tendency of the technology to encourage *anticipatory war* (preventive or preemptive war), which is war initiated to avoid a perceived future threat from one's opponent. For states with the requisite cyber technology, it may seem so easy and tempting to initiate war in a conflict situation that the result would be that the requirement of last resort is effectively ignored. There are several reasons for this. First, an operational cyberattack may seem like such an obvious thing to undertake when a state perceives a future threat from its opponent, given that such an attack promises severe disruption of the opponent's military capability without a great deal of physical destruction. Second, such a disruption provides the attacking state with a great military advantage, perhaps an effective decapitation by itself forcing the opponent's surrender. Third, a state may believe that its operational cyberattack, because it would cause little physical damage, would not even be considered an act of war. I have argued that this is an illusion, given that severe military disruption must count as an act of war, but it may be an illusion to which states are prone. Moreover, even if an initial cyberattack was short of an act of war, the likelihood of escalation to war would be very great. I will return to this issue in the next section.

Now consider cyberattacks in the context of the other aspect of just war theory, *jus in bello*, the morality of how a war is fought. Again, there are a set of criteria, in this case *discrimination*, *proportionality*, and *due care*. Discrimination requires that attacks in war be directed against military targets rather than civilian targets. Proportionality requires that attacks of war be such that the contribution they are expected to make to victory in the war outweighs the expected amount of harm they would do. (*In bello* proportionality differs from *ad bellum* proportionality in that it is applies to individual military actions rather than to the war as a whole and does not assume a just cause). Due care requires that attacks in war minimize expected harm to civilians. Some argue that it would be easier to keep cyberwar within the limits defined by these criteria than to keep conventional war so limited, that cyberattacks "if rightly handled, could end up being more discriminate, more proportional, and thus more in compliance with… the moral principles of *jus in bello*, than any conventional counterpart" (Lucas 2010, p. 297). But this judgment is hasty. Cyberattacks raise some problems with each of these criteria.

Consider discrimination. This criterion would, of course, rule out strategic cyberattacks, as it rules out strategic conventional attacks. Military objects can be deliberately attacked, but civilian objects cannot. Civilian infrastructure cannot be made the object of military attack. But can a clear line be drawn between military and civilian objects? This raises the problem of so-called dual-use infrastructure, infrastructure that serves both military and civilian purposes, such as electrical power grids. This is also a problem in the case of conventional attack. For example, the United States has taken a permissive view on what dual-use infrastructure it may attack. In the first Gulf War, it treated the electric power grid of Iraq as liable to attack. But these attacks resulted in the deaths of an estimated seventy to ninety thousand civilians (Bayles 2001, p. 52). In the indirect deaths of the civilians were taken into account, this seems like a strategic attack rather than an operational attack (though the issue of intentionality would complicate this judgment).[13] Although this problem arises in conventional war (as it did in the Gulf War), it is a special problem in the case of cyberattack because infrastructure is a natural point of attack in cyberspace (Hirschland 2001, p. 11). Infrastructure is to an increasing extent under the control of computer systems.

This dual-use problem is exacerbated in the case of cyberattack because such attack seems benign in comparison with a conventional attack. Destroying the electrical grid of a nation with conventional weapons, even precise ones, would likely kill hundreds of civilian power workers, while to do so with a cyberattack may kill

[13] For an argument against an understanding of the rules of war that would allow such a permissive view of the liability of dual-use infrastructure to attack, see (Shue and Wippman 2002).

no one directly. This is part of the illusion that cyberattack is a bloodless strategy, which was discussed in the first section. It is an illusion, as the facts of the Gulf War illustrate. One way to think of the illusion is this. Consider that the purpose of traditional kinetic strikes against enemy combatants is not directly to kill them, but rather to disable them, to make them unable to resist one's own forces (this is the basis of the rule of war protecting injured combatants from attack). It just so happens that with present technology the only effective way to disable combatants is usually to kill them. The illusion is that cyber technology seems to promise a way to disable the opponent as a whole by destroying infrastructure without killing anyone. It is an illusion because the destruction of the infrastructure will lead to large numbers of civilian deaths.

In addition, there are two other special problems for the criterion of discrimination posed by cyberattacks. First is the problem of the combatant status crucial to the application of discrimination. The criterion assumes that there is a clear distinction between combatants and civilians, but cyber technology muddies the distinction due to "the use of typically civilian technology and know-how to conduct military operations via computer" (Schmitt 2002, p. 398). Discrimination becomes more difficult to apply because many civilians will be intimately involved in the activity of war. Second there is the problem of *perfidy*. Deception is a recognized part of war, and most deception, referred to as *ruse*, is permissible, but some deception, perfidy, is not acceptable. One example of perfidy is the feigning of a status protected under the rules of war, such as combatants pretending to be civilians. This is a violation of discrimination. Cyber technology offers great opportunity for deception in general and perfidy in particular. For example, a state could, in an act of perfidy, plant an "all clear" message into the opponent's communications systems just before an attack (Bayles 2001, p. 50). Cyber technology would increase the opportunity and temptation for states to engage in perfidy.

The *in bello* criterion of proportionality works in tandem with the criterion of discrimination under the moral framework know as the *doctrine of double effect*. The idea is that while discrimination precludes attacks intended to harm civilians, some expected civilian harm, if not intended, may be permissible, just in case it satisfies proportionality. This sort of moral calculation arises especially in the case of cyber technology because, as mentioned, the natural targets of cyberattack are the computer systems controlling the infrastructure on which civilians depend on for survival. Here the illusion that cyberattacks are bloodless again plays a role, leading those applying the doctrine of double effect to tend to ignore the long-term harm to civilians from infrastructure attacks. In addition, there is the problem mentioned earlier in discussion of *ad bellum* proportionality that the uncertain expectations about the effects of a cyberattack could lead the attackers to launch a more devastating attack to insure that it has the desired effects. Added to this is the potential for what are called "reverberating effects," which is the tendency, due to interconnectivity, for effects in one realm or region to produce effects in another, often in a completely unpredictable way (Schmitt 1998–1999, pp. 893–894). All of these points show how the criterion of proportionality would be more difficult to satisfy in the case of cyberattacks.

But as a counterweight to these concerns, there is, also as mentioned earlier, a way in which the criterion of proportionality may be more easily satisfied through the use of cyber technology. While cyberattacks can easily impose great costs on civilians, even when these costs are not intended or even foreseen, they also may potentially be used in a way that imposes lesser civilian costs than corresponding conventional attacks would do. The difference lies in the way cyberattacks and conventional attacks do the damage they do. In the case of an opponent's electrical power grid, for example, destruction through conventional attacks would mean that the grid would be out of commission for weeks or months, if not longer, given the need to rebuild it. But a cyberattack could probably be designed to knock out the grid only temporarily, given that it could be done without any destruction of the facilities. The destruction might be done in a way that was effectively reversible (National Research Council 2009, p. 264). Indeed, cyberattacks "may make it possible to achieve desired military aims with less collateral damage and incidental injury than in traditional kinetic attacks" (Schmitt 2002, p. 397).

This potential difference in the minimal amount of destruction with which cyberattack and conventional attack can be carried out has an important bearing on the relevance of cyber technology to the third *in bello* criterion, due care. Due care requires that an attack be done in the way that minimizes the amount of civilian harm, and for a state with the relevant technological capability, this will usually be through a cyberattack. Michael Schmitt suggests that "military commanders will in certain cases be obligated to employ their cyber assets in lieu of kinetic weapons when collateral and incidental effects can be limited" (Schmitt 2002, pp. 397–398). Following the demands of the due care criterion, the acquisition of cyber capability and its use in preference to conventional attack may become morally obligatory!

In summary, there are some difficulties (and also some advantages) that cyber technology potentially poses for the application of the criteria of *jus in bello*. But the difficulties seem not to be sufficient to find that the technology threatens to make just war theory irrelevant. Many of these are connected with the illusion of bloodlessness, and this is something that combatants and military leaders can be educated to reject. Cyber war is not a new kind of war, in the sense that it requires different moral rules about how it is fought. A similar judgment seems appropriate for the criteria of *jus ad bellum*, with one important exception. For all of the *ad bellum* criteria save one, the difficulties we have considered that arise when they are applied to cyberattacks are not sufficient to find that the technology threatens to make just war theory irrelevant. The one exception is the criterion of last resort. A case could be made that cyber technology would make this criterion inapplicable, at least in practice, threatening the relevance of just war theory to cyberattack. Whether this is in fact the case is considered in the next section.[14]

---

[14] Of course, I must note that the judgments made in this section, as with other of the judgments in this essay, are at least partly speculative, given that the future development of the technology and the way it turns out to be applied in the real world of war cannot be accurately predicted.

7.3 Section III

Speaking of the implications of the military use of cyber technology, Arquilla notes that "the one area that may change is the use of force in preventive ways" (Arquilla 1999, p. 387). The greatest problem that cyber technology poses for just war theory is the potential that this technology could lead to a great increase in anticipatory wars, thereby vitiating the likelihood that war initiation through cyberattack could satisfy the *ad bellum* criterion of last resort. Anticipatory war is a war where the belligerent strikes the first blow while justifying the attack on the defensive grounds that the attack is in response to an expected attack from the opponent. Among anticipatory wars are preventive wars and preemptive wars. Preventive wars occur when the attacker believes that its opponent intends to strike at some indefinite time in the future. Preemptive wars occur when the attacker strikes first in reasonable fear that it opponent's own first strike is imminent. In both sorts of case, the attacker believes that if it lands the first blow it is more likely to win the war it expects sooner or later. Preventive wars always and preemptive wars sometimes violate the criterion of last resort because they ignore the peaceful alternatives that may be available to avoid war.

In order to draw some conclusions about the relevance of last resort to cyber conflict, we need to make a brief excursion into strategic thought as it applies to cyberattack. One of the reasons that anticipatory war is a special problem in the context of cyber technology is that cyberattacks are potentially very effective in an effort to "prepare the battle space," that is, to weaken an opponent in preparation for a conventional attack. Preparatory cyberattacks could provide a great advantage in a conventional war by disrupting the opponent's military communications, its intelligence gathering assets, its global positioning systems, and so forth (Schmitt 1998–1999, p. 929). The capacity of cyber technology to achieve such results is one reason why with this technology, as with nuclear weapons technology, the offense dominates the defense (Nye 2011, p. 21). It is easier to destroy assets with cyberattack than to protect them from cyberattack. All of these factors lead to a situation of *crisis instability*, a situation in which war is more likely to break out in a crisis because both sides have incentives to initiate an attack (National Research Council 2009, p. 306). The upshot is that cyber technology "makes war more thinkable." (Arquilla 1999, p. 398). George Lucas raises the concern that cyber technology will "lower the threshold for resorting to war of any sort, traditionally consigned to being the last (rather than the earliest) resort to conflict resolution with adversaries or competitors" (Lucas 2010, p. 294). War becomes more thinkable, and the last resort criterion is devalued or ignored.

These points may be developed by our considering some comparisons between the strategic implications of cyber and nuclear technology. While the two technologies differ dramatically in the amount of destruction they can cause, there are comparisons in the strategic environment each creates[15] (National Research Council

[15] Joseph Nye notes that a strategic cyberattack could send the economy back to 1990, while a strategic nuclear attack could sent the economy back to the Stone Age (Nye 2011, p. 22).

2009, p. 295). Both represent a dominance of the offense over the defense. Consider first how offense dominance works in case of nuclear technology. First, compare conventional deterrence and nuclear deterrence. In the case of conventional deterrence, a state's effort to deter its opponent is based on the threat both to inflict costs and to deny benefits, on a threat of *punishment* and on a threat of *denial*. Denial is the ability of a state's defensive capabilities to blunt the success of an attack. So, even if deterrence fails, the success of an attack is not guaranteed. But in the case of nuclear weapons, defenses are ineffective because there is no adequate way to stop a large number of nuclear warheads on missiles from getting through to their targets. There is no denial and nuclear deterrence rests exclusively on the threat of retaliatory punishment. It might seem that this would lead to great crisis instability, as there would be great advantage to going first, but this turned out not to be the case. The reason is that both sides were able to proliferate and protect warheads, making their retaliatory capacity partly invulnerable to surprise attack; each side was guaranteed to have enough warheads left over after a surprise attack to destroy the attacker. Each side had the capacity for assured destruction, and together the United States and the Soviet Union were in a state of MAD, mutual assured destruction. There was no advantage and so no incentives for going first because whichever side went first, both sides would be destroyed.

Now consider some analogies (and disanalogies) between these features of nuclear strategy and the potential strategic environment of cyber technology.[16] Consider first offense dominance. This does not mean quite the same thing for cyber technology as it does for nuclear technology. In the case of nuclear technology offense dominance is due to the destructive power of the offense and the impossibility of effective defense. In the case of cyber technology offense is not as destructive and defenses may have some effectiveness. On the side of defenses, offense dominance is due to uncertainty about how effective cyber defenses would be and to the greater cost of defense as compared with offense. As Randall Dipert notes, cyber defense is unlikely to be sufficiently successful and likely to be too expensive (Dipert 2010, p. 403). On the side of offenses, offense dominance is due to disruptive effects of operational cyberattacks and the social destruction possible with strategic cyberattacks. But the social destruction of cyberattack is much more tolerable, much less of a punishment, than that of nuclear attack. What is more to the point, the advantage of operational cyberattack comes mainly from going first, creating incentives to strike first that may not be outweighed by the threat of punishment. In the case of nuclear deterrence, in contrast, the threat of punishment far outweighs the advantages from going first.

Consider other features of cyber deterrence. (I understand cyber deterrence to mean deterrence *of* a cyber attack, whether *by* cyber threats or conventional threats). There are some features of cyberattack that make cyber deterrence less credible, hence less effective. The most important of these is the attribution problem, dis-

[16] This discussion largely concerns adversarial relations between "near-peer" states, those roughly equal in military capability. Different factors may arise in the relations between adversaries in an asymmetrical power relationship.

cussed earlier. A state that believes that it can attack anonymously because its attack will not be successfully attributed to it will not be deterred by a threat of retaliation. Even if attribution can be achieved in most cases, it will likely take time, and threats of delayed retaliation are less credible than threats of immediate retaliation. In contrast with the nuclear threat, "the difficulties of attack attribution leave a comparable [cyber] threat with far less credibility"[17] (National Research Council 2009, p. 2, 294, 295). At the same time, the fact that cyber defenses could have some effectiveness means that cyber deterrence (unlike nuclear deterrence) has an element of denial, increasing, to that extent, its credibility (Nye 2011, pp. 33–34).

Using cyber threats to deter cyberattacks may not be an effective form of deterrence due, among other reasons, to uncertainty about how effective cyber retaliation would be. (Again, in contrast, there is little uncertainty about the effects of a nuclear retaliation). For this reason, conventional threats (or even nuclear threats) may be a necessary part of an effective posture of deterrence of cyberattacks. This seems to be current U.S. policy, as the Whitehouse has declared: "When warranted, the United States will respond to hostile acts in cyberspace [reserving] the right to use all necessary means" (Whitehouse 2011, p. 14). But including conventional threats to deter cyberattack would pose problems of its own. A conventional response to cyberattack, Arquilla remarks, "may tend toward escalation" (Arquilla 1999, p. 390). The reason is that such a response would be perceived as upping the ante. This perception may result from two features of the situation. First, the amount of harm done by the initial cyberattack may not be clear, not only to the attacker but to the victim, and second, a comparison between harm done in a cyber attack and harm done in a retaliatory conventional attack is inherently difficult to make (Arquilla 1999, p. 391). Given the bias each side has toward its own case, these two features could lead to a perception on the part of the recipient of the retaliatory attack that that attacker has upped the ante, thereby calling for the recipient to up the ante further in response. This means that throwing conventional attacks into a military exchange begun with cyberattacks would make the *signaling* necessary to avoid escalation more difficult (National Research Council 2009, p. 308). An escalatory spiral could easily result. This sort of situation suggests a systemic weakness of cyber deterrence. The situation has a dilemmatic structure: cyber deterrence can be restricted to cyber threats or can include conventional threats as well; if the former, the deterrence posture is weak, and if the latter it is weak as well.

So cyber deterrence is weak due to the attribution problem, and it is weak due to the sort of dilemmatic structure just noted. When deterrence is weak, the likelihood of one side or the other initiating a cyberattack is greater. This contributes to instability, where the likelihood of one side initiating a cyberattack rises even further. If side A recognizes that side B is more likely to initiate a cyberattack (because deterrence is recognized to be weak), side A itself becomes more likely to initiate a

[17] Another factor in the need for a delay before the retaliatory response is that it may take time for the victim of a cyberattack to figure out how much damage was done, which it needs to know before it can decide how great the retaliation should be (or even whether it should occur at all) (National Research Council 2009, p. 310).

cyberattack out of fear that B might do so, which then leads to B being more likely to do so, and so forth. This dynamic has been called in the case of nuclear strategy the reciprocal fear of surprise attack. Surprise attack would be an anticipatory war. This creates crisis instability because a crisis in the relationship between A and B is the time when this dynamic is most likely to engage. In the nuclear context, this dynamic is forestalled by the prospects of mutual destruction, but this prospect is not available to forestall the dynamic in the case of cyberattack.

Now we may return from our excursion into cyber strategy to the discussion of *jus ad bellum* and the criterion of last resort. Anticipatory war is generally a violation of last resort, always in the case of preventive war and often in the case of preemptive war. The maturation of cyber military technology will increase the risk of anticipatory war, due to the weakness of cyber deterrence, along with the tendency of potential belligerents to treat a cyberattack as less than a full-fledged act of war, compounded by the fact that there is a deep inherent advantage to going first, due in part to the way in which initial cyberattack works to "prepare the battle space" for a more general war. Due to crisis instability, the pressure on states to initiate cyberattack will sometimes be great, and this pressure means that the last resort criterion will often be ignored. War becomes more thinkable because the last resort criterion is not being thought about.

The earlier arguments have shown that, in prospect, cyber technology make the other *ad bellum* criteria and the *in bello* criteria more difficult, but not impossible to adhere to. But the dynamics of cyberattack and cyber deterrence may show that the last resort criterion is, as a matter of practice, impossible to adhere to. The fear that each side has, especially in a crisis, that the other is about to attack will make it often impossible for either side to effectively explore options short of war for resolving the conflict. In this sense, cyber military technology makes this criterion of last resort irrelevant, and to this extent the maturation of cyber military technology would take us beyond just war theory.

References

Arquilla, John. 1999. Ethics and information in warfare. In *The changing role of information in warfare,* ed. Z. Khalilzad et al. 379–401. Santa Monica: Rand Corporation.

Bayles, William. 2001. The ethics of computer network attack. *Parameters* 31:44–58.

Clarke, Richard, and Robert Knake. 2010. *Cyber war: The next threat to national security and what to do about it.* New York: Harper Collins.

Dilpert, Randall. 2010. The ethics of cyberwarfare. *The Journal of Military Ethics* 9:384–410.

Hirschland, Matthew. 2001. Information warfare and the new challenges to waging just war presented at conference of the American Political Science Association, Denver, CO.

Lucas, George R. 2011. Permissible preventive cyber warfare, Proceedings of the Air Force Research Institute on the Future of Cyber Power, ed. Pano Yanakageorgos, et al.

Lucas, George R. 2010. Postmodern war. *Journal of Military Ethics* 9:289–298.

National Research Council. 2009. *Technology, policy, law, and ethics regarding U.S. acquisition and use of cyberattack capabilities.* Washington, D.C.: National Academies Press.

Nye, Joseph. 2011. Nuclear lessons for cybers. *Strategic Studies Quarterly* 5 (4):18–38.

Powell, Michael. 1998. The deaths he cannot sanction; Ex-U.N. Worker details harm to Iraqi children. *Washington Post*, 17 December.

Rid, Thomas. 2011. Cyber war will not take place. *Journal of Strategic Studies* 35:5–32.

Rowe, Neil. 2010. The ethics of cyberweapons in warfare. *International Journal of Cyberethics* 1:20–31.

Sanger, David. 2012. Obama order sped up wave of cyberattacks against Iran. *New York Times*, 1 June.

Schmitt, Michael. 1998–1999. Computer network attack and the use of force in international law: Thoughts on a normative framework. *Columbia Journal of Transnational Law* 37:885–937.

Schmitt, Michael. 2002. Wired warfare: Computer network attack and *Jus in Bello*. *International Review of the Red Cross* 84:365–398.

Shue, Henry, and David Wippman. 2002. Limiting attacks on dual-use facilities performing indispensable civilian functions. *Cornell International Law Journal* 35:559–577.

Whitehouse. 2011. International strategy for cyberspace. http://www.whitehouse.gov/sites/default/files/rss_viewer/international_strategy_for_cyberspace.pdf.

Chapter 8
Information Warfare and Just War Theory

Mariarosaria Taddeo

Abstract This article is devoted to developing an ethical analysis of information warfare, the warfare waged in the cyber domain. It has the twofold goal of filling the theoretical vacuum surrounding this phenomenon and of providing the grounding for the definition of new ethical regulations for information warfare. The article maintains that Just War Theory is a necessary but not sufficient instrument for considering the ethical implications of information warfare and argues that a suitable ethical analysis of this kind of warfare is developed when Just War theory is merged with Information Ethics. The initial part of the article describes information warfare and its main features, and highlights the problems that arise when Just War Theory is endorsed as a means of addressing ethical problems engendered by information warfare. The final part introduces the main aspects of Information Ethics and defines three principles for a just information warfare.

8.1 Introduction

The cyberspace is nowadays conceived as the fifth domain in which war may be waged, along with land, sea, air and space, for the ability to control, disrupt or manipulate the enemy's informational infrastructure has become as decisive with respect to the outcome of conflicts as weapon superiority. In this respect, information and communication technologies (ICTs) have proved to be a useful and convenient technology for waging war.

The military deployment of ICTs has radically changed the way wars are declared and waged nowadays. It has actually determined the latest revolution in military affairs, i.e. the informational turn in military affairs (Toffler and Toffler

M. Taddeo (✉)
Cyber Security & Ethics, Department of Politics and International Studies,
University of Warwick, Coventry, United Kingdom
e-mail: M.Taddeo@warwick.ac.uk

Uehiro Centre for Practical Ethics, University of Oxford, Oxford, United Kingdom

L. Floridi, M. Taddeo (eds.), *The Ethics of Information Warfare,*
Law, Governance and Technology Series 14, DOI 10.1007/978-3-319-04135-3_8,

1997).[1] Such a revolution is not the exclusive concern of the military; it has also a bearing on ethicists and policymakers, since existing ethical theories of war and national and international regulations struggle to address the novelties of this phenomenon.

This article is devoted to developing an ethical analysis of information warfare (IW). It has the twofold goal of filling the theoretical vacuum surrounding this phenomenon and of providing the conceptual grounding for the definition of new ethical regulations for IW. The proposed analysis rests on the conceptual investigation of IW provided in (Taddeo 2012), which highlights the informational nature of this phenomenon, and argues that IW represents a profound novelty, which reshapes the very concept of war and raises the need for new ethical guidelines.

On the basis of this analysis, the article maintains that Just War Theory (JWT) is a necessary but not sufficient instrument for considering the ethical implications of IW. It is argued that investigating IW through the lens of JWT allows for the unveiling of fundamental ethical issues that this phenomenon brings to the fore, yet that attempting to address these issues solely on the basis of this theory will leave them unsolved.

It is suggested that problems encountered when addressing IW through JWT are overcome if the latter is merged with Information Ethics (Floridi 2013). This is a macro-ethical theory, which is particularly suitable for taking into account the features and the ethical implications of *informational phenomena*, like internet neutrality (Turilli et al. 2012), online trust (Turilli et al. 2010), peer-to-peer (Taddeo and Vaccaro 2011) and IW.

Merging the principles of JWT with the macro-ethical framework provided by Information Ethics has two advantages; it allows the development of an ethical analysis of IW capable of taking into account the peculiarities and the novelty of this phenomenon; and it also extends the validity of JWT to a new kind of warfare, which at first glance seemed to fall outside its scope (Taddeo 2012).

The initial part of this article will describe IW and its main features. It will then focus on JWT and on the problems that arise when this theory is endorsed as a means of addressing the case for IW. Information Ethics will then be introduced. Its four principles will provide the grounds for the analysis proposed in the final part of this article, where the principles for a just IW are defined. Finally, it is discussed how JWT can be applied to IW without leading to ethical conundrums. Having delineated the path ahead, we should now begin our analysis by considering in more detail the nature of IW.

8.2 Information Warfare

The expression 'information warfare' has already been used in some parts of the extant literature to refer solely to the uses of ICTs devoted to breaching the opponent's informational infrastructure in order to either disrupt it or acquire relevant data and

[1] For an analysis of revolution in military affairs considering both the history of such revolutions and the effects of the development of the most recent technologies on warfare see (Benbow 2004; Blackmore 2005).

Fig. 8.1 The different uses of ICTs in military strategies. (Taddeo 2012)

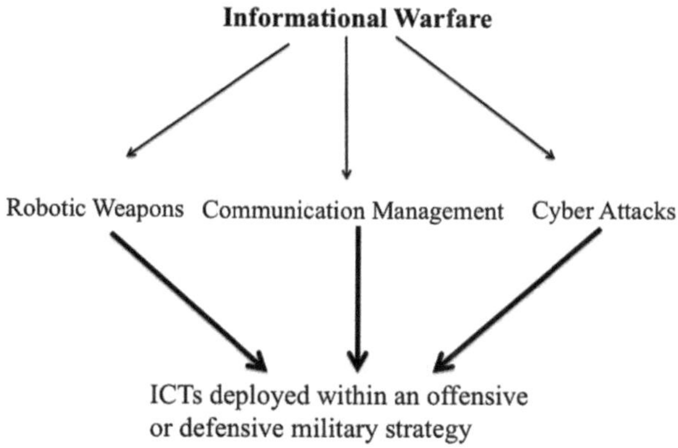

information about the opponent's resources, military strategies and so on; see for example (Libicki 1996; Waltz 1998; Schwartau 1994).

The distributed denials of service (DDoS) attacks conducted in 2007 against institutional Estonian websites, the attacks launched to block the Internet communication in Burma during the 2010 elections[2] or the injection of Stuxnet, a computer-worm in the Iranian nuclear facilities of Bushehr [3] provide good examples of how ICTs can be used to conduct so-called cyber attacks. Cyber attacks are surely one of the most well-known and debated forms of ICT-based conflicts, but they should be considered only one form of IW. Equating IW with cyber attacks would lead to a too restrictive use of the label IW.

In the rest of this article, IW will refer to a wide spectrum of phenomena, encompassing cyber-attacks as well as the deployment of robotic-weapons and ICT-based communication protocols (see Fig. 8.1).

The reason for endorsing such a wide spectrum definition is twofold. On the one side, it allows for focusing on the purpose for the military deployment of ICTs rather than on the mode of their deployment. In the case of IW, the endorsement of ICTs—be it the use of (semi)autonomous weapons, of a computer virus, or of digital devices to enhance the performance of forces on the battlefield—has a *disruptive* intent. Such an intent is the main concern of the ethical analysis proposed in this article. On the other side, endorsing a wide spectrum definition has also methodological advantage. For by considering indiscriminately the different uses of ICTs in warfare, the analysis provides ethical principles addressing the totality of the cases of IW rather than some of its specific occurrences.

A parallel with the ethical analysis of traditional warfare will support such a methodological choice. JWT is concerned with warfare in general, its principles are valid in any theatre of traditional warfare, be it waged with swords or guns or by deploying nuclear weapons as long as the weapons are used with the same intent, namely to inflict physical damage on the enemy. Likewise, the analysis proposed

[2] http://www.bbc.co.uk/news/technology-11693214 http://news.bbc.co.uk/2/hi/europe/6665145.stm.

[3] http://www.cbsnews.com/stories/2010/11/29/world/main7100197.shtml.

in this article aims to provide ethical principles for a just IW valid for every mode of conducting it.

This approach neither undermines the differences between the use of a computer virus and a robotic weapon nor denies that such different uses generate different ethical issues. Rather, it asks the reader to be patient and to focus first on the aspects that are common among the different military uses of ICTs, since the analysis of these aspects provides the groundwork for addressing specific ethical problems brought to the fore by specific military uses of ICTs.

Following this approach, IW is defined as follows:

> **Information Warfare** is the use of ICTs within an offensive or defensive military strategy endorsed by a state and aiming at the immediate disruption or control of the enemy's resources, and which is waged within the informational environment, with agents and targets ranging both on the physical and non-physical domains and whose level of violence may vary upon circumstances. (Taddeo 2012)

This definition highlights two aspects of IW: its *informational nature* and its *transversality*. The informational nature of IW is a consequence of the fact that this kind of warfare rests on the military deployment of technological artefacts devoted to elaborating, managing and communicating data and information. In this respect, IW shows how it is related to the so-called Information revolution.

The Information revolution is a multi-faceted phenomenon. It rests on the development and the ubiquitous dissemination of the use of ICTs, which have a wide impact on many of our daily practises: from working and interacting with other human beings, to driving and planning holidays. ICTs allow for developing and acting in a new domain, the digital or informational one (Floridi 2009). This is a completely virtual, non-physical domain, which has grown important and hosts a considerable relevant part of our lives. With the information revolution we witness a shift, which has brought the *non-physical domain* to the fore and made it as important and valuable as the physical one (Taddeo 2012).

IW is one of the most compelling instances of such a shift. It shows that there is a new environment, where physical and non-physical entities coexist and are equally valuable, and in which states have to prove their authority and new modes of warfare are being developed specifically for deployment in such a new environment.[4]

The shift toward the non-physical domain provides the ground for the transversality of IW. This is a complex aspect that can be better understood when IW is compared with traditional forms of warfare.

Traditionally, war entails the use of a state's *violence* through the state *military* forces to determine the conditions of governance over a determined territory (Gelven 1994). It is a necessarily violent phenomenon, which implies the sacrifice of human lives and damage to both military and civilian infrastructures. The problem

---

[4] The USA only spent $ 400 million in developing technologies for cyber conflicts: http://www.wired.com/dangerroom/2010/05/cyberwar-cassandras-get-400-million-in-conflict-cash/.

The UK devoted £ 650 million to the same purpose: http://www.theinquirer.net/inquirer/news/1896098/british-military-spend-gbp650-million-cyber-warfare.

to be faced when waging traditional warfare is how to minimise damage and losses while ensuring the enemy is overpowered.

IW is different from traditional warfare in several respects. It is not a necessarily violent and destructive phenomenon (Arquilla 1998). For example, IW may involve a computer virus capable of disrupting or denying access to the enemy's database, and in so doing it may cause severe damage to the opponent without exerting *physical* force or violence. In the same way, IW does not necessarily involve human beings. An action of war in this context can be conducted by an autonomous robot, such as, for example, the EADS Barracuda, and the Northrop Grumman X-47B,[5] or by an autonomous cruising computer virus (Abiola et al. 2004), targeting other artificial agents or informational infrastructures, like a database or a website. IW can be waged exclusively in a digital context without ever involving concrete targets. Nevertheless, IW may escalate to more violent forms. Consider for example the consequences of a cyber attack targeting a military aerial control system causing aircraft to crash (Waltz 1998).

As remarked above, the transversality of IW is the key feature of this phenomenon; it is the aspect that differentiates it the most from traditional warfare. Transversality is also the feature that engenders the ethical problems posed by IW. The potential bloodless and non-destructive nature of IW (Denning 2009; Arquilla 1998) makes it desirable from both an ethical and a political perspective, since at first glance, it seems to avoid bloodshed and it liberates political authority from the burden of justifying military actions to the public. A more attentive analysis unveils that IW can lead to highly violent and destructive consequences, which would be dangerous for both military forces and civil society. For this reason declaring and waging IW requires strict regulation to guarantee its fairness.

To this end an analysis that discloses the ethical issues that IW engenders and points at the direction for their solution is a preliminary and necessary step. The development of such analysis will be the task of the next section.

8.3 IW and Just War Theory

Ethical analyses of war are developed following three main paradigms: JWT, Pacifism or Realism. In the rest of this paper, the analysis will focus only on JWT. Two reasons support this choice: (i) the ethical problems with which JWT is concerned are generated by the very same decision to declare and to wage war, be it a traditional or an informational war. Therefore JWT sheds light on the analysis of the ethical issues posed by IW; (ii) The criteria for a *just* war proposed by this theory remain valid when considering IW, for the justification to resort to war and the cri-

[5] Note that MQ-1 Predators and EADS Barracuda, and the Northrop Grumman X-47B are Unmanned *Combat* Aerial Vehicles used for combat actions and they are different from Unmanned Air Vehicles, like for example Northrop Grumman MQ-8 Fire Scout, which are used for patrolling and recognition purposes only.

teria for *jus in bello* and *post bellum* proposed by JWT rest on the defence of basic human rights of life and liberty, see for example (Walzer 2000). There is no doubt that such rights and their preservation hold in the case of traditional warfare as well as in the case of IW.

Nevertheless, despite the relevance of these two reasons, it would be mistaken to consider JWT both the necessary and sufficient ethical framework for the analysis of IW, for addressing this new form of warfare solely on the basis of JWT generates more ethical conundrums than it solves. In the words of Arquilla (Arquilla 1998):

> it appears that [...] information war [has] left a good part of 'just war theory' in tatters. For IW may now make preventive war far more thinkable (and practical), straining the limits of the concept of 'right purpose'. And the manner in which the information revolution empowers small groups and individuals to wage IW suggests that the notion of 'duly constituted authority' may also have lost meaning. Finally, the ease in undertaking IW operations, and the fact that they are disruptive, but not very destructive, weakens notions of justice as requiring that war be started only as a 'last resort'. (p. 208)

The ethical problems encountered when addressing IW on the basis of JWT originate from the differences between IW and traditional warfare. Such differences need to be taken into account in developing an ethical analysis of IW. Otherwise, the risk is twofold. On the one side, if the peculiarities of IW are not taken in consideration one is caused to disregard all those cases of IW that do not correspond to the parameters of traditional warfare (mainly the non-violent cases of IW). These are nevertheless potentially dangerous cases and need to be regulated as they remain disruptive and may cause extensive damage. On the other side, not taking into account the novelty posed by IW and focusing only on traditional criteria when analysing this phenomenon leads to a focus only on those cases which fall within the scope of traditional warfare, namely the violent cases of IW. In this case, the ethical analysis equates these instances of IW to traditional warfare, and leaves unexplored the peculiarities of IW and its specific ethical implications.

Particularly relevant in considering the differences between traditional and informational warfare is the transversality of the ontological status of the entities involved in the latter. Traditional warfare concerns human beings and physical objects, while IW involves *artificial* and *non-physical* entities alongside human beings and physical objects. Therefore, there is a hiatus between the ontology of the entities involved in traditional warfare and of those involved in IW. Such a hiatus affects the ethical analysis, for JWT rests on an anthropocentric ontology, i.e. it is concerned with respect for *human* rights and disregards all non-human entities as part of the moral discourse, and for this reason it does not provide sufficient means for addressing the case for IW (more details on this aspect presently).

The case of the autonomous cruising computer virus will help in clarifying the problems at stake (Abiola et al. 2004). These viruses are able to navigate through the web and identify autonomously their targets and attack them without requiring any supervision. The targets are chosen on the basis of parameters that the designers encode in the virus, so there is a boundary to the autonomy of these agents. Still, once the target has been identified the virus attacks without having to receive 'authorisation' from the designer or any human agent.

In considering the moral scenario in which the virus is launched three main questions arise. The first question revolves around the identification of the moral agents, for it is unclear whether the virus itself should be considered the moral agent, or whether such a role should be attributed to the designer or to the agency that decided to deploy the virus, or even to the person who actually launched it. The second question focuses on moral patients. The issue arises as to whether the attacked computer system itself should be considered the moral receiver of the action, or whether the computer system and its users should be considered the moral patients. Finally, the third questions concerns the rights that should be defended in the case of a cyber attack. In this case, the problem is whether any rights should be attributed to the informational infrastructures or to the system compounded by the informational infrastructure and the users.

These questions indicate that IW includes informational infrastructures, computer systems, and databases. In doing so, it brings new objects into the moral discourse. The first step toward an ethical analysis of IW is to determine the moral status of such (informational) objects and their rights. Help in this respect is provided by Information Ethics, which will be introduced in the Sect. 4. Before focusing on Information Ethics, we shall first consider in detail some of the problems encountered when applying three principles of JWT to IW.

8.3.1 The Tenets of JWT and IW

For the purpose of this analysis, we shall consider whether and how the tenets of *last resort*, *more good than harm*, and *non-combatants immunity* can be applied in the case of IW.

The principle of 'war as last resort' prescribes that a state may resort to war only if it has exhausted all plausible, peaceful alternatives to resolve the conflict in question, in particular diplomatic negotiations. This principle rests on the assumption that war is a violent and sanguinary phenomenon and as such it has to be avoided until it remains the only reasonable way for a state to defend itself. The application of this principle is shaken when IW is taken in consideration, because in this case war may be bloodless and may not involve physical violence at all. In these circumstances, the use of the principle of war as last resort becomes less immediate.

Imagine, for example, the case of tense relations between two states and that the tension could be resolved if one of the states decide to launch a cyber attack on the other state's informational infrastructure. The attack would be bloodless as it would affect only the informational grid of the other state and there would be no casualties. The attack could also lead to resolution of the tension and avert the possibility of a traditional war in the foreseeable future. Nevertheless, according to JWT, the attack would be an act of war, and as such it is forbidden as a first strike move.

The impasse is quite dramatic, for if the state decides not to launch the cyber attack it will be probably forced to engage in a sanguinary war in the future, but if the state authorises the cyber attack it will breach the principle of war as last resort and commit an unethical action.

This example is emblematic of the problems encountered in the attempt to establish ethical guidelines for IW. In this case, the main problem is due to the transversality of the modes of combat described in Sect. 2, which makes it difficult to define unequivocal ethical guidelines. In the light of the principle of last resort, soft and non-violent cases of IW can be approved as means for avoiding traditional war (Perry 1995), as they can be considered a viable alternative to bloodshed, which may be justly endorsed to avoid traditional warfare (Bok 1978). At the same time, even the soft cases of IW have a disruptive purpose—disrupting the enemy's (informational) resources (Arquilla and Ronfeldt 1997). Such a disruptive intent, even when it is not achieved through violent and sanguinary means, must be taken in consideration by any analysis aiming at providing ethical guidelines for IW.[6]

Another problem arises when considering the principle of 'more good than harm'. According to this principle, before declaring war a state must consider the *universal* goods expected to follow from the decision to wage war, against the *universal* evils expected to result, namely the casualties that the war is likely to determine. The state is justified in declaring war only when the goods are proportional to the evils. This is a fine balance, which is straightforwardly assessed in the case of traditional warfare, where evil is mainly considered in terms of casualties and physical damages which may result from a war. The equilibrium between the goods and the evils becomes more problematic to calculate when IW is taken into consideration.

As the reader may recall, IW is transversal with respect to the level of violence. If strictly applied to the non-violent instances of IW, the principle of more good than harm leads to problematic consequences. For it may be argued that, since IW can lead to the victory over the enemy without determining casualties, it is a kind of warfare (or at least the soft, non-violent instances of IW) that is always morally justified, as the good to be achieved will always be greater than the evil that could potentially be caused.

Nonetheless, IW may result in unethical actions—destroying a database with rare and important historical information, for example. If the only criteria for the assessment of harm in warfare scenarios remain the consideration of the physical damage caused by war, then an unwelcome consequence follows for all the non-

[6] It is worthwhile noticing that the problem engendered by the application of the principle of last resort to the soft-cases of IW may also be addressed by stressing that these cases do not fall within the scope of JWT as they may be considered cases of espionage rather than cases of war, and as such they do not represent a 'first strike' and the principle of last resort should not be applied to them. One consequence of this approach is that JWT would address war scenarios by focusing on traditional cases of warfare, such as physical attacks, and on the deployment of robotic weapons, disregarding the use of cyber attacks. This would be quite a problematic consequence because, despite the academic distinction between IW and traditional warfare, the two phenomena are actually not so distinct in reality. Robotic weapons fight on the battlefield side by side with human soldiers, and military strategies comprise both physical and cyber attacks. By disregarding cyber attacks, JWT would be able to address only partially contemporary warfare, while it should take into consideration the whole range of phenomena related to war waging in order to address the ethical issues posed by it (for a more in depth analysis of this aspect see (Taddeo 2012)).

violent cases of IW comply by default to this principle. Therefore, destroying a digital resource containing important records is deemed to be an ethical action, as it does not constitute physical damage *per se*.

The problem that arose with the application of this principle to the case of IW does not concern the validity *in se* of the principle. It is rather the framework in which the principle has been provided that becomes problematic. In this case, it is not the prescription that the goods should be greater than the harm in order to justify the decision to conduct a war, but rather is the set of criteria endorsed to assess the good and the harm that shows its inadequacy when considering IW.

A similar problem arises when considering the principle of 'discrimination and non-combatant immunity'. This principle refers to a classic war scenario and aims at reducing bloodshed, prohibiting any form of violence against non-combatants, like civilians. It is part of the *jus in bello* criteria and states that soldiers can use their weapons to target exclusively those who are "engaged in harm" (Walzer 2000, p. 82). Casualties inflicted on non-combatants are excused only if they are a consequence of a non-deliberate act. This principle is of paramount importance, as it prevents massacres of individuals not actively involved in the conflict. Its correctness is not questionable yet its application is quite difficult in the context of IW.

In classic warfare, the distinction between combatants and non-combatants reflects the distinction between military and civil society. In the last century, the spread of terrorism and guerrilla warfare weakened the association between non-combatants and civilians. In the case of IW such association becomes even feebler, due to the blurring between civil society and military organisations. (Schmitt 1999; Shulman 1999; Taddeo 2012).

The blurring of the distinction between military and civil society leads to the involvement of civilians in war actions and raises a problem concerning the discrimination itself: in the IW scenario it is difficult to distinguish combatants from non-combatants. Wearing a uniform or being deployed on the battlefield are no longer sufficient criteria to identify someone's social status. Civilians may take part in a combat action from the comfort of their homes, while carrying on with their civilian life and hiding their status as informational warriors.

It would be misleading to consider the problems described in this sections as reasons for dismissing JWT when analysing IW. These problems rather point to a more fundamental problem; namely the need to consider more carefully the case of IW, and to take into account its peculiarities.

8.4 Information Ethics

The time has come to introduce Information Ethics. This is a macro-ethics, which is concerned with the whole realm of reality and provides an analysis of ethical issues by endorsing an informational perspective. Such an approach rests on the consideration that "ICTs, by radically changing the informational context in which moral issues arise, not only add interesting new dimensions to old problems, but lead us to rethink, methodologically, the very grounds on which our ethical positions are based" (Floridi 2006, p. 23).

In one sentence Information Ethics is defined as a *patient-oriented*, *ontocentric*, and *ecological* macroethics. Information Ethics is patient-oriented because it considers the morality of an action with respect to its effects on the receiver of the action. It is ontocentric, for it endorses a non-anthropocentric approach for the ethical analysis. It attributes a moral value to all the existing entities (both physical and non-physical) by applying the principle of ontological equality: "This ontological equality principle means that any form of reality […], simply for the fact of being what it is, enjoys a minimal, initial, *overridable*, equal right to exist and develop in a way which is appropriate to its nature" (Floridi 2013). The principle of ontological equality is grounded on an information-based ontology, according to which all existing things can be considered from an informational standpoint and are understood as informational entities, all sharing the same informational nature.

By endorsing such a principle, Information Ethics guarantees a judgment of the moral scenario free from a biological or anthropological bias, for, following the principle of ontological equality, minimal and overridable rights to exist and flourish pertain to all existing things, and not just to human or living things. From this perspective, the Colosseum, Jane Austin's writings, a human being and computer software all share the right to exist and flourish, as they are all informational entities.[7]

A clarification is now necessary to avoid any misunderstanding. Information Ethics endorses a minimalist approach, it considers informational nature as the minimal common denominator among all existing things. Such a minimalist approach should not be mistaken for reductionism, as Information Ethics does not claim that informational ontology is the unique perspective from which moral discourse is addressed. Rather it maintains that the informational perspective provides a minimal starting point, which can then be enriched by considering other moral perspectives.

In this respect, it is worthwhile emphasising that the principle of ontological equality does not imply that all entities have the same moral value. The rights attributed to the entities are *initial*, they are overridden whenever they conflict with the rights of other (more morally valuable) entities. The moral value of an entity is determined according to its potential contribution to the enrichment and the flourishing of the informational environment. Such an environment, the *Infosphere*, includes all existing things, be they digital or analogical, physical or non-physical and the relations occurring among them, and between them and the environment. The blooming of the Infosphere is the ultimate good, while its corruption, or destruction, is the ultimate evil.

In particular, any form of corruption, depletion and destruction of informational entities or of the Infosphere is referred to as *entropy*. Lest the reader be confused, in this case entropy refers to "any kind of *destruction* or *corruption* of informational

[7] For more details on the information-based ontology see (Floridi 2003). The reader interested in the debate on the Informational ontology and the principles of Information Ethics may whish to see (Floridi 2007).

objects (mind, not of information), that is, any form of impoverishment of *being*, including *nothingness*, to phrase it more metaphysically", (Floridi 2013) and has nothing to do with the concept developed in physics or in information theory (Floridi 2007).

Information Ethics considers the duty of any moral agent with respect to its contribution to the informational environment, and considers any action that affects the environment by corrupting or damaging it, or by damaging the informational objects existing in it, as an occurrence of entropy, and therefore as an instance of evil (Floridi and Sanders 1999, 2001). On the basis of this approach Information Ethics provides four principles to identify right and wrong and the moral duties of an agent. The four moral principles are:

0. entropy ought not to be caused in the infosphere (null law);
1. entropy ought to be prevented in the infosphere;
2. entropy ought to be removed from the infosphere;
3. the flourishing of informational entities as well as of the whole infosphere ought to be promoted by preserving, cultivating and enriching their properties.

These four principles together with the theoretical framework of Information Ethics will provide the ground to proceed further in our analysis, and define the principles for a just IW.

8.5 Just IW

The first step toward the definition of the principles for a just IW is to understand the moral scenario determined by this phenomenon. The framework provided by Information Ethics proves to be useful in this regard, for we can now answer the questions posed in Sect. 3 concerning the identification of moral agents, moral patients and the rights that have to be respected in the case of IW. The remainder of this article will focus on the problems regarding moral patients and their rights. The issue concerning the identification of moral agents in IW requires an in-depth analysis (see for example (Asaro 2008)) which falls outside the scope of this article. I shall clarify a few aspects concerning morality of artificial agents relevant to the scope of this analysis, before setting this issue aside.

The debate on morality of artificial agents is usually associated to the issues of ascribing to artificial agents moral responsibility for their actions. (Floridi and Sanders 2004) provide a different approach to this problem decoupling the moral *accountability* of an artificial agent, i.e. its ability to perform morally qualifiable actions, from the moral *responsibility* for the actions that such an agent may perform.

Floridi and Sanders argue that an action is morally qualifiable when it as morally qualifiable effects on its patient, and that every entity that qualifies as an interactive, autonomous and adapTable (transition) system and which performs a morally qualifiable action is (independently from its ontological nature) considered a morally accountable agent. So when considering the case for IW, a robotic weapon and a computer virus are considered moral agents as long as they show some degree of autonomy in interacting and adapting to the environment and perform actions that may cause either moral good or moral evil.

As argued by Floridi and Sanders, attributing moral accountability to artificial agents extends the scope of ethical analysis to the actions of such agents and permits prescribable moral principles for their actions. This approach particularly suits the purpose of the present analysis, for the reader may accept suspending judgment on the moral responsibility for the actions that artificial agents may perform in case of IW, and agree that such actions are nevertheless morally qualifiable, and that as such they should be the objects of a prescriptive analysis.

Once we have put aside the issue concerning the morality of artificial agents, we are left with questions concerning the moral stance of the receivers of the actions performed by such agents and of the rights that ought to be respected in the case of IW. The principle of ontological equality states that all (informational) entities enjoy some minimal rights to exist and flourish in the Infosphere, and therefore every entity deserves some minimal respect, in the sense of a "disinterested, appreciative and careful attention" (Hepburn 1984; Floridi 2013).

When applied to IW, this principle allows for considering all entities that may be affected by an action of war as moral patients. A human being, who enjoys the consequences of a cyber attack and an informational infrastructure that is disrupted by a cyber attack are both to be held moral patients, as they are both the receivers of the moral action. Following Information Ethics, the moral value of such an action is to be assessed on the basis of its effects on the patients' rights to exist and flourish, and ultimately on the flourishing of the Infosphere.

The issue then arises concerning which and whose rights should be preserved in case of IW. The answer to this question follows from the rationale of Information Ethics, according to which an entity may lose its rights to exist and flourish when it comes into conflict (causes entropy) with the rights of other entities or with the well-being of the Infosphere. It is a moral duty of the other inhabitants of the Infosphere to *remove* such a malicious entity from the Infosphere or at least to impede it from perpetrating more evil.

This framework lays the ground for the first principle for just IW. The principle prescribes the condition under which the choice to resort to IW is morally justified.

I. IW ought to be waged only against those entities that endanger or disrupt the well-being of the Infosphere.

Two more principles regulate just IW, they are:

II. IW ought to be waged to preserve the well-being of the Infosphere.
III. IW ought not to be waged to promote the well-being of the Infosphere.

The second principle limits the task of IW to restoring the *status quo* in the Infosphere before the malicious entity began increasing the entropy within it. IW is just as long its goal is to *repair* the Infosphere from the damage caused by the malicious entity.

The second principle can be described using an analogy; namely, IW should fulfil the same role as police forces in a democratic state. It should act only when a crime has been, or is about to be, perpetrated. Police forces do not act in order to ameliorate the aesthetics of cities or the fairness of a state's laws; they only focus on reducing or preventing crimes from being committed. Likewise, IW ought to be endorsed as an *active* measure in response to increasing of evil and not as proactive strategy to foster the flourishing of the Infosphere. Indeed, this is explicitly forbidden by the third principle, which prescribes the promotion of the well-being of the Infosphere as an activity that falls beyond the scope of a just IW.

These three principles rest on the identification of the moral good with the flourishing of the Infosphere and the moral evil with the increasing of entropy in it. They endorse an informational ontology, which allows for including in the moral discourse both non-living and non-physical entities. The principles also prescribe respect for the rights of such entities along with those of human beings and other living things, and respect for the rights of the Infosphere as the most fundamental requirement for declaring and waging a just IW.

In doing so the three principles overcome the ontological hiatus described in Sect. 3, and provide the framework for applying JWT to the case of IW without leading to the ethical conundrums analysed in Sect. 3.1. The description of how JWT is merged with Information Ethics is the task of the next section.

8.6 Three Principles for a Just IW

The application of the principle of 'last resort' provides the first instance of the merging of JWT and Information Ethics. The reader may recall that the principles forbids embracing IW as an 'early move' even in those circumstances in which IW may avert the possibility of waging a traditional war. The principle takes into account traditional (violent) forms of warfare, and it is coupled with the principle of 'right cause', which justifies resort to war only in case of 'self-defence'. However right this approach may be when applied to traditional (violent) forms of warfare, it proves inadequate when IW is taken into consideration. The impasse is overcome when considering the principles for just IW.

The first principle prescribes that any entity that endangers or disrupts the well-being of the Infosphere loses its basic rights and becomes a licit target. The second principle prescribes that a state is within its rights to wage IW to re-establish the *status quo* in the Infosphere and to repair the damage caused by a malicious entity. These two principles allow for breaking the deadlock described in Sect. 3.1, because a state can rightly endorse IW as an early move to avoid the possibility of a traditional warfare, as the latter threatens greater disruption of the Infosphere, and as such it is deemed to be a greater evil (source of entropy) than IW.

A caveat must be stressed in this case: the waging of IW must comply with the principles of 'proportionality' and 'more good than harm'. In waging IW, the endorsed means must be sufficient to stop the malicious entity, and in doing so the means ought not to generate more entropy than a state is aiming to remove from the

Infosphere in the first place. This leads us to consider in more detail the principle of 'more good than harm'.

The issues that arose in the case of IW are due to the definition of the criteria for the assessment of the 'good' and the 'harm' that a warfare may cause. As described in Sect. 3.1, endorsing traditional criteria leads to a serious ethical conundrum, since all (the majority of) the cases of IW that do not target physical infrastructures or human life comply by default to this principle regardless of their consequences.

Such a problem is avoided if damage to non-physical entities in considered as well as physical damage. More precisely, the assessment of the good and the harm should be determined by considering the general condition of the Infosphere 'before and after' waging the war. A just war never determines greater entropy than that in the Infosphere before it was waged. Once considered from this perspective, the principle of more good than harm acts as corollary of the second principle for just IW. It ensures that a just IW is waged to restore the *status quo* and does not increase the level of entropy in the Infosphere.

Increasing entropy in the Infosphere also provides a criterion for reconsidering the application of the principle of 'discrimination and non-combatants' immunity' to IW. As it has been argued in Sect. 3.1, IW blurs the distinction between militaries and civilians, as it neither requires military skills nor does it require a military status of the combatants to be waged. This makes problematic the application of this principle to IW; nevertheless the principle has to be maintained as it prescribes the distinction between licit and illicit war targets.

Help in applying this principle to IW comes from the first principle for just IW, which allows for dispensing with the distinction between militaries and civilians, and for substituting it with the distinction between licit targets and illicit ones. The former are those malicious entities who endanger or disrupt the well-being of the Infosphere. According to the principle, IW rightfully targets only malicious entities, be they military or civilian. The social status ceases to be significant in this context, because any entity that contributes to increasing the evil in the Infosphere loses its initial rights to exist and flourish and therefore becomes a licit target. More explicitly, it becomes a moral duty for the other entities in the Infosphere to prevent such entity from causing more evil.

Before concluding this article, I shall briefly clarify an aspect of the proposed analysis, lest the reader be tempted to consider it warmongering.

The third principle provided in Sect. 5 stresses that IW is never justly waged when the goal is improving the well-being of the Infosphere. This principle rests on the very same rationale that inspires Information Ethics, according to which the flourishing of the Infosphere is determined by the blooming of informational entities, of their relations and by their well-being. IW is understood as a form of disruption and as such, by definition, it can never be a vehicle for fostering the prosperity of the Infosphere nor is it deemed to be desirable *per se*. IW is rather considered a necessary evil, the bitter medicine, which one needs to take to fight something even more undesirable, i.e. the uncontrolled increasing of the entropy in the environment. With this clarification in mind we can now pull together the threads of the analysis proposed in this article.

8.7 Conclusion

The goal of this article is to fill the conceptual vacuum surrounding IW and of providing the ethical principles for a just IW. It has been argued that to this purpose JWT provides the necessary but not sufficient tools. For although its ideal of just warfare grounded on respect for basic human rights in the theatre of war holds also in the case of IW, it does not take into account the moral stance of non-human and non-physical entities which are involved and mainly affected by IW.

This article defends the thesis that in order to be applied to the case for IW, JWT needs to extend the scope of the moral scenario to include non-physical and non-human agents and patients. Information Ethics has been introduced as a suitable ethical framework capable of considering human and artificial, physical and non-physical entities in the moral discourse. It has been argued that the ethical analysis of IW is possible when JWT is merged with Information Ethics. In other words, JWT *per se* is too large a sieve to filter the issues posed by IW. Yet, when combined with Information Ethics, JWT acquires the necessary granularity to address the issues posed by this form of warfare.

The first part of this paper introduces IW and analyses its relation to the information revolution and its main feature, namely its transversality. It then describes the reasons why JWT is an insufficient tool with which to address the ethical problems engendered by IW and continues by introducing Information Ethics. The second part of the article defends the thesis according to which once the ontological hiatus between the JWT and IW it is bridged, JWT can be endorsed to address the ethical problems posed by IW.

The argument is made that such a hiatus is filled when JWT encounters Information Ethics, since its ontocentric approach and informational ontology allow for ascribing a moral status to any existing entity. In doing so, Information Ethics extends the scope of the moral discourse to all entities involved in IW and provides a new ground for JWT, allowing it to be extended to the case for IW.

In concluding this article I should like to remark that the proposed ethical analysis should in no way be understood as a way of advocating warfare or IW. Rather it is devoted to prescribing ethical principles such that if IW has to be waged then it will at least be a just warfare.

References

Abiola, A., J. M. Munoz, and W. J. Buchanan. 2004. *Analysis and detection of cruising computer viruses*. Paper presented at the 3rd International Conference on Electronic Warfare and Security.

Arquilla, J. 1998. Can information warfare ever be just? *Ethics and Information Technology* 1:203–212.

Arquilla, J., and D. Ronfeldt. 1997. *In Athena's camp: Preparing for conflict in the information age*. Santa Monica: RAND Corporation.

Asaro, P. 2008. How just could a robot war be? In *Current issues in computing and philosophy,* eds. P. Brey, A. Briggle, and K. Waelbers, 50–64. Amsterdam: IOS Press.

Benbow, T. 2004. *The magic bullet?: Understanding the revolution in military affairs*. London: Brassey.

Blackmore, T. 2005. *War X*. Toronto: University of Toronto Press Incorporated.

Bok, S. 1978. *Lying: Moral Choice in Public and Private*. New York, USA: Pantheon.

Denning, D. E. 2009. The ethics of cyber conflict. In *The handbook of information and computer ethics*, eds. K. E. Himma and H. T. Tavani, 407–428. New York: Wiley.

Floridi, L. 2003. On the intrinsic value of information objects and the infosphere. *Ethics and Information Technology* 4:287–304.

Floridi, L. 2006. Information ethics, its nature and scope. *SIGCAS Computer and Society* 36:21–36.

Floridi, L. 2007. Understanding information ethics. *APA Newsletter On Philosophy and Computers* 7:3–12.

Floridi, L. 2009. The information society and its philosophy. *The Information Society* 25:153–158.

Floridi, L. 2013. *Information ethics*. Oxford University Press.

Floridi, L., and J. W. Sanders. 1999. Entropy as evil in information ethics. *Etica & Politica* 1: special issue on Computer Ethics I(2).

Floridi, L., and J. W. Sanders. 2001. Artificial evil and the foundation of computer ethics. *Ethics and Information Technology* 3:55–66.

Floridi, L., and J. W. Sanders. 2004. On the morality of artificial agents. *Minds and Machines* 14:349–379.

Gelven, M. 1994. *War and existence*. Philadelphia: Pennsylvania State University Press.

Hepburn, R. 1984. *Wonder and other essays*. Edinburgh: Edinburgh University Press.

Libicki, M. 1996. *What is information warfare*? Washington, DC: National Defense University Press.

Perry, David L. 1995. *Repugnant Philosophy: Ethics, Espionage, and Covert Action*. Journal of Conflict Studies, Springer.

Schmitt, M. N. 1999. The principle of discrimination in 21st century warfare. *Yale Humna Right and Development Law Journal* 2:143–160.

Schwartau, W. 1994. *Information warfare: Chaos on the electronic superhighway*. New York: Thunder's Mouth Press.

Shulman, M. R. 1999. Discrimination in the laws of information warfare. *Columbia Journal of Transnational Law* 37:939–968 (Pace Law Faculty Publications).

Taddeo, M. 2012. Information warfare: A philosophical perspective. *Philosophy and Technology* 25:105–120.

Taddeo, M., and A. Vaccaro. 2011. Analyzing peer-to-peer technology using information ethics. *The Information Society* 27:105–112.

Toffler, A., and H. Toffler. 1997. Foreword: The new intangibles. In *In Athena's camp: Preparing for conflict in the information age*, eds. J. Arquilla and D. Ronfeldt, xii–xxiv. Santa Monica: RAND.

Turilli, M., A. Vaccaro, and M. Taddeo. 2010. The case of on-line trust. *Knoweldge, Technology and Policy* 23:333–345.

Turilli, M., A. Vaccaro, and M. Taddeo. 2012. Internet neutrality: Ethical issues in the internet environment. *Philosophy & Technology* 25:133–151.

Waltz, E. L. 1998. *Information warfare principles and operations*. Norwood: Publisher Artech House, Inc.

Walzer, M. 2000. *Just and unjust wars: A moral argument with historical illustrations*. 3rd ed. New York: Basic Books.

Part III
Information warfare and virtue

Chapter 9
The Wrong in Cyberattacks

Thomas W. Simpson

Abstract Cyberattacks raise tricky and important moral questions. When is it permissible to attack an opponent's Information and Computing Technology systems (ICTs)? Or obligatory? Could a cyberattack be a legitimate *casus belli*, being justifiably responded to by physical force, or are the only justifiable responses those that are in-kind? Existing discussion of the moral framework with which to evaluate the ethics of cyberattacks has consisted either in the application of Just War theory, or in claims that Just War theory is not fit-for-purpose for twenty-first Century warfare and that a new kind of ethical theory is required. Both sides are partly right and partly wrong. Advocates of Just War theory are correct that it applies to cyberattacks, but are wrong to suppose that this constitutes a sufficient basis on which morally to evaluate all cyberattacks. The dissenters are correct that an alternative framework is required for the ethical evaluation of cyberattacks, but they are wrong to suppose it must be novel. This judgment is justified by my substantive thesis: we already have a concept which is fit for purpose in evaluating a great proportion of the moral significance of cyberattacks, namely that of *harm to property*. Property rights provide an ethical framework within which cyberattacks should be assessed. An important advantage of such an approach is that it makes sense of how cyberattacks can be both bloodless but constitute real harm. A result is that a cyberattack could be a legitimate *casus belli*, given sufficient harm to property that was sufficiently (morally) valuable.

9.1 Introduction

By subverting systems which use information and computing technologies (ICTs), enemies can damage or commandeer critical infrastructure; steal secrets; disable or seize control of weaponry; empty bank accounts; and render us unable to com-

T. W. Simpson (✉)
Blavatnik School of Government, University of Oxford, Oxford, UK
e-mail: thomas.simpson@bsg.ox.ac.uk

L. Floridi, M. Taddeo (eds.), *The Ethics of Information Warfare*,
Law, Governance and Technology Series 14, DOI 10.1007/978-3-319-04135-3_9,

municate. They may do so with just a computer screen and an Internet connection. *A fortiori*, they may do so without any explosions and without crossing any borders in person.

It is an aggressive and hostile act to subvert someone's ICTs. That is why they are commonly called 'cyber*attacks*'. Strategic questions about cyberattacks arise: is it wise for me to engage in cyberattacks, given the ways an enemy might respond? How should I react if attacked? Moral questions also arise: am I permitted to prosecute cyberattacks against an enemy? Or even sometimes obliged? Given a cyberattack against me, what response is justified?

These questions are relatively new because our reliance on ICTs is relatively new. But reflection on moral issues raised by the use of force is hardly novel. Just War theory has long recognised the use of force as a practical necessity in this vale of tears, whilst denying that pragmatic considerations on whether and how it should be used are the only relevant ones. There are also moral constraints. While exponents of Just War theory differ in the details, there is substantial overlap in the general approach, to the degree that a tradition can be recognised. It states conditions which must be satisfied for a decision to go to war to be just and for a war to be conducted justly, termed the *jus ad bellum* and *jus in bello* conditions respectively.

It is therefore natural that initial discussions of the ethics of force in cyberspace have enquired into whether and how such acts can satisfy the *jus ad bellum* and *jus in bello* conditions (Arquilla 1999, 2010; Cook 2010; Rowe 2007, 2008, 2010). Dissent has followed swiftly, with claims that Just War theory is not fit for purpose in the moral evaluation of cyberwarfare (Dipert 2010) and with proposals of alternative and novel ethical frameworks (Taddeo 2012). This last is an example of what Herman Tavani calls the 'uniqueness thesis', viz. that some ethical issues raised by ICTs cannot be adequately accommodated by traditional categories of moral discourse (2002, 2005).

Who is right? On what basis should the morality of act-types and act-tokens of cyberattacks be assessed—the criteria given by Just War theory, or some other? My answer is: both are partly right. Advocates of Just War theory are correct that it applies to cyberattacks, but are wrong to suppose that this constitutes a sufficient basis on which to evaluate them morally. The dissenters are thus correct that an alternative framework is required for the ethical evaluation of cyberattacks, but they are wrong to suppose it must be a novel one. This judgment is justified by my substantive thesis: we already have a concept which is fit for purpose in evaluating a great proportion of the moral significance of cyberattacks, namely that of *harm to property*. Property rights provide an ethical framework within which cyberattacks should be assessed. I thereby deny the uniqueness thesis and explain why Just War theory is appropriate for judgment about the permissibility of cyberattacks in some contexts but not all.

My thesis is 'traditionalist' in Tavani's terms. Traditionalists claim that ethical issues raised by ICTs can be well accommodated by established categories of moral discourse. The defence of traditionalism is necessarily piecemeal. For each morally-evaluable practice created by a new technology, the traditionalist must show how established categories of moral discourse apply to it. In showing that cyberattacks give no reason to deny traditionalism, I contribute to the more general defence of that position.

The structure of the paper is as follows. I first show the insufficiency of Just War theory as a framework for evaluating the moral permissibility of cyberattacks (§ 9.2). I then defend my central thesis, that the notion of property provides this framework. I explain the central terms of my proposal (§ 9.3); identify its implications regarding how one should come to a practical judgement about the justice of a particular cyberattack (§ 9.4); and then justify the proposal (§ 9.5). I conclude by replying to objections (§ 9.6).

9.2 The Limits of War

While some cyberattacks are acts of war, not all are. It depends on context. In consequence, Just War theory is the wrong place to start in thinking about the ethics of cyberattacks. Nonetheless, non-war cyberattacks are not exempt from moral consideration. The intention to cause harm and the harm potentially or actually caused by a cyberattack is sufficient to show this. To substantiate these claims, consider the following actual cyberattacks.

In 'Operation Orchard' in September 2007, a few Israeli jets crossed into Syrian airspace and destroyed what is widely believed to have been a nuclear reactor under construction at the Dair Alzour site (BBC 2007; for an identification of the function of the site, see IAEA 2011). The aircraft used were F-15 and F-16 fighter-bombers. These are not equipped with stealth technology. Yet Syrian air defence systems failed to detect the incursion of the fighters until too late. It is widely speculated, by US Air Force officials among others, that part of the reason for the Israelis' success was their use of a programme similar to that developed by BAE Systems named 'Suter'. Suter allows its operators to override the controls of ground-based air defence systems and direct sensors away from approaching aircraft so that no alarms are sounded (Leyden 2007; Gasparre 2008). Operation Orchard was an act of war and very likely had a cyberattack as an integral part.

A more recent attack was less clearly an act of war. The computer virus Stuxnet was first identified in June 2010. It is a highly engineered piece of malware which targets only a specific type of Siemens system, namely those which run the 'Step 7' software application. It is spread by infected USB sticks and once on a computer replicates itself to others on an internal network. 'Its final goal is to reprogram industrial control systems (ICS) by modifying code on programmable logic controllers (PLCs) to make them work in a manner the attacker intended and to hide those changes from the operator of the equipment' (Falliere et al. 2011, p. 1). These Siemens systems are integral to the presumed Iranian nuclear weapons programme. Stuxnet's effects are not publicly known, but Meir Dagan, the former head of Mossad, is taken to have alluded to them when he told the Knesset in January 2011 that 'a series of setbacks' meant that Iran was unlikely to acquire a bomb before 2015 (Melman 2011). The complexity of the malware is such that a nation state was likely behind it, with Israel and the US the most likely candidates.

The writing and distribution of Stuxnet was an aggressive and hostile act against Iran. But in terms of moral evaluation, it was more akin to an intelligence operation than an act of war. Its distribution required nothing more overt than non-uniformed personnel. They need not have carried weapons for self-defence. As it is customary to allocate responsibility for particular kinds of acts to different institutions, the widely shared belief that the CIA played some role in the attack—and not US CYBERCOM, the Pentagon's 4-star command for cyberwarfare—is further support for the claim, albeit not conclusive.

Here is a cyberattack which was no act of war. In April 2012, a group of hackers calling themselves 'TeaMp0isoN' claimed to have bombarded the Metropolitan Police's Anti-Terrorism hotline with over 700 bogus calls. This was to distract officers from a security breach which had enabled the group to listen in on internal conversations among the officers (Furness 2012). Two teenagers have since been arrested, while Scotland Yard deny that the recordings were of internal conversations and that there had therefore been a security breach (BBC 2012b). While anarchists may apply the label 'act of war' to Team Poison's actions in a literal sense, for the rest of us it is only a euphemism. The authorities' response also suggests this. It is the police who are dealing with the cyberattack, not the military.

A final case: James Jeffrey was convicted in April 2012 of stealing the records of 10,000 women who were registered on the website of the British Pregnancy Advisory Service (BPAS). Reportedly an occasional member of the hacker collective 'Anonymous', he did so by accessing the BPAS site remotely (BBC 2012a). Although he did not publish the details of the women Jeffrey boasted about the hack on Twitter, an action the police found helpful when trying to identify the culprit. BPAS is a private association, so the case shows that cyberattacks need not be targeted at only state institutions. While Jeffrey's actions were criminal, they were not an act of war.

The foregoing cases illustrate that whether a cyberattack is an act of war depends on the context. Thomas Rid gives a plethora of examples which further illustrate the point (2011). It depends on who carried it out; whether they did so in a public or private capacity; what the function of the system is which is targeted; and who the targeted system belongs to. It should not be surprising that context is key in determining whether a cyberattack is an act of war. The same is true of physical violence. Consider setting fire to a vehicle. This was an act of war if the fire-raiser was a member of the Resistance in France in 1943 bombing a Wehrmacht staff car. It was merely arson if they were a disenchanted youth in the 2005 riots in the Parisian *banlieues* enjoying the sight of the Citroens of the bourgeoisie going up.

It is because moral judgements are sensitive to social context that Just War theory is the wrong place to start in thinking about the ethics of cyberattacks. Just War theory has proven a robust and useful framework for thinking about the ethics of when and how to use force during war. It does not claim to provide a framework for evaluating when and how to use force generally. Not all force occurs during war. Force is sometimes used during espionage. It is used by the police. Force is used by gangs in turf wars over drug patches. It occurs in the playground. Just War theory is silent on how morally to evaluate force in these contexts; they lie outside its scope. Nonetheless, we are able to argue about whether force is legitimate in these contexts. This is because we possess a range of moral concepts which are not restricted to contexts of war but are generally applicable.

This point applies to the debate over the ethics of cyberattacks as follows. While cyberattacks are morally evaluable, not all cyberattacks are acts of war. So Just War theory is not a sufficient framework for the ethical assessment of all cyberattacks. Thus extending the International Law of Armed Conflict to non-state actors involved in non-war cyber activities is as absurd as extending it to cover the use of force by gangs in inner-city Manchester (against Denning 2008).

A different starting point is required. It is a desideratum on the ethical framework with which cyberattacks are morally evaluated that it be applicable to non-war forms of cyber-conflict as well as those which occur in war. It is a further desideratum that whatever this framework is, it also explains how Just War theory can be applicable to some cyberattacks. I turn now to identify and defend a framework which satisfies these desiderata.

9.3 ICTs as Property

The notion of *property*, and its cognates *property rights* and *harm to property*, provide a basis for judging whether a significant class of cyberattacks are just. In this section I explain the assertion. In the next I outline its implications. In § 5 I justify it.

A basic social practice is the ascription of ownership over physical objects or spaces. When someone owns an object or a space, that object or space is their *property*. The conditions under which someone comes to own property are culturally variable. But I know of no anthropological report of a culture which has no institution of property. (Communism is not a counter-example. Even if a society has existed where no individual has owned anything—which is dubious in the extreme—property is still owned collectively). Ownership of property confers *property rights*. These consist in the entitlement of the property owner to dispose of her property as she wishes (without breaching others' rights). Property rights also oblige others not to interfere with her property. If someone interferes with her property, two wrongs may be done. First, they wrong her by broaching her property rights. Second, they may wrong her by reducing the value of her property. Call this latter, *harm to property*. The two are different; it is possible to harm the owner of some property by infringing her rights even if no damage is done. The trespasser harms the owner by breaching her rights against people crossing her land without permission.

A principal function of property rights is to ensure that harm to property does not occur and that owners of property thereby maintain the value of that property. The justification of property rights is a nice philosophical question going back at least to John Locke. For present purposes I assume that they exist and are justified.

The second kind of wrong done when property rights are breached, namely harm to property, is my interest here. Harm to property comes in degrees, because the value of property comes in degrees. In ensuring that harm to property does not occur, part of the purpose of property rights is to protect the value of property. Property may have value in at least the following three ways. It has *financial value*, namely how much money was paid to acquire the property or would have to be paid to replace it. Property can also have value through *attachment*. This may be due to its emotional or historical significance, for instance, and need not necessarily be reflected by its financial value. This is most obvious with personal items—lockets of hair, photographs, family heirlooms and so on. While a financial value can be given for these by finding out how much the owner would be prepared to pay for it, there is a clear sense in which such items are irreplaceable. Take parents who hold precious the 'if I should die' letter given to them when their soldier son was fatally wounded in Afghanistan. Although they can insure against its loss through fire, the money that would be paid out is only a substitute for, not a replacement of, its real value.

Another way that property can have a non-financial value is through its *practical value*. The practical value of property is the non-financial value that is secured by that property, and is shown by the consequences of its being interfered with. I think here of such basic values as life and health. Practical value usually correlates with financial value, but not always. Consider a specific bit of hardware—a cog in an engine, for instance, which is very specifically engineered for the job it is to perform and without which the engine cannot perform its function. Suppose that the engine performs a vital task, such as providing power to a hospital in a place where there is no alternative electricity source. Suppose further it is hard to get hold of replacements for the cog because of its rarity. The cog is extremely important to the functioning of the hospital even though it is inexpensive. Therefore practical value does not entail financial value. Nor does financial value entail practical value. This is shown by the fashion industry.

Property can hold value because of the function it fulfils in a system. This is a variant of practical value, and the hospital generator example illustrates this point also. The cog is significant not so much 'in itself', but rather because it is necessary for the generator to function. The generator is a mechanical system. The generator is further embedded in mechanical/electrical system, namely the electricity circuit in the hospital on which a plethora of equipment relies. This equipment in turn fulfils functions in the larger system of the hospital, a system which includes doctors, nurses, administrators and so on. The sum function of the hospital is to treat and ameliorate illness and injury. The value of some property is a function of how reliant on that item all 'higher' systems are, and what the practical effect is achieved by the higher systems.

Organisations own property and this property may have value in any of the above ways. Think of St Paul's Cathedral and the Crown Jewels for the UK, or the Statue of Liberty and the Declaration of Independence for the USA. These are examples of property owned by collective agents which are principally valued in terms of attachment. The hospital's electricity generator is an example of an item of property owned by a collective agent which is principally valued in terms of practical value.

Property rights are *prima facie*, not *ultima facie*, in W. D. Ross's terms (Ross 1930). They are permissibly overridden by competing rights which are of greater value. The right to life trumps someone's property rights, for instance. In fleeing an attacker intent on killing me, I may permissibly run across someone else's land. It is beyond scope here to list those kinds of rights which may trump property rights, were that even possible. The present point is that we readily make judgments about the relative weights of different rights.

I now apply these general considerations to the specific issue of ICTs. Computers and other information technologies can be property. I own the PC on my desk. The Ministry of Defence owns those on theirs. And so on. ICTs are valuable in at least the above three ways; they cost money; people grow attached to their computer and to the informational content that it makes available; and they are practically important not least in terms of the systems which depend on them to function properly.

Cyberattacks breach property rights. When an enemy attacks my ICTs and changes the way they function, they interfere with my property without my permission. In changing the way my computer works or stopping it working, the attacker generally causes it to lose financial value; they may destroy or damage something which the system supports which has attachment value; and there may be practical consequences arising from the system's failure. Restoring the value of damaged ICTs takes time, work and money. So cyberattacks cause harm to property—property which may be owned by individuals or collective agents.

Part of the degree of harm caused by a cyberattack is the value of that which an ICT system enables. As a dramatic example, consider pacemakers which regulate a patient's heartbeat and have some basic computing technology to relay information without surgery. Halperin et al. (2008) show that some pacemakers are vulnerable to remote interference. A person's life depends on that property not being interfered with. Harm to that property, through a cyberattack, is thus extremely serious.

9.4 Moral Implications

Property rights provide an ethical framework for judging whether a cyberattack is just. Someone is not justified in broaching my property rights over my ICT solely because they are able to. If someone interferes with my PC without my permission and causes damage or destroys its value, the harm to my property that the offender has caused makes them liable to rectifying justice. They may be punished and required to restore the lost value of my property.

There is one exception to the above, namely when my property rights are permissibly overridden. It requires a good reason for my rights to be overridden. I have not attempted to delineate what other rights may justify property rights being breached. But note the following. How weighty my property rights are judged to be varies with how valuable the property at issue is. The more valuable the property, the more important must be the competing rights if they are rightly to override my property rights. Because the value of some property can be considered only in the particular, so the weight of property rights must be addressed afresh on each occasion. So the more important the function fulfilled by a particular computer or ICT system, the more serious a cyberattack against that system is.

Determining that I hold property rights over the ICTs I own provides only a starting point for adjudging whether a particular past or putative cyberattack is just. The starting point is the following question: given how weighty my property rights are over this particular ICT systems, do others have competing rights which are weightier? Only if one were able to enumerate in advance all the competing rights which may permissibly override property rights would it be possible to provide more than this. Nonetheless, we are capable of making judgements about the relative weights of competing sets of rights in practical situations. So the given starting point is of practical utility.

The ethical framework provided by the notion of property rights applies to non-war contexts. Property rights may permissibly be breached by the police, for instance. Police permissibly break into and enter someone's house when they have a search warrant. Addicts looking for TVs to sell for drug money do so impermissibly. Similarly for cyberattacks. The procedural framework for controlling how official agencies subvert others' ICT systems is less well developed. But we are nonetheless able to judge what the right kinds and the wrong kinds of reasons are. Because the framework provided by property rights applies to cyberattacks in non-war contexts, it satisfies the first desideratum noted above (§ 2).

The same framework explains how Just War theory is applicable to cyberattacks. Questions of *jus in bello* are straightforward. Soldiers are permitted to breach property rights in prosecuting a war because there are competing rights which frequently enough are weightier than property rights, viz. their rights to self-preservation and self-defence. According to the *jus in bello* criteria, soldiers prosecute a war justly only if they fight in a way which is *discriminatory* and *proportional* (see Guthrie and Quinlan 2007, pp. 35–43; Coady 2008, pp. 107–131). These necessary conditions are met often enough in conventional, 'kinetic' war, as the military term it. They can also be met by cyberattacks. Cyberattacks are capable of targeting combatants only and of being a proportional use of force; the use of Suter in Operation Orchard is an example. (Although note Betz and Stevens' argument that cyberattacks are seldom so discriminatory; 2011, pp. 39–42). So cyberattacks can be just in the context of a war.

Questions of *jus ad bellum* are harder. Arguably the hardest one is the following: is a conventional counterattack ever morally justified by an enemy cyberattack? Equivalently, is a cyberattack a legitimate *casus belli*? Dipert calls this a 'hard case' (2010, p. 392), and I agree with the characterisation. I will defend an answer using the framework provided by property rights. In doing so, I also provide a diagnosis of why the case is hard. Such a diagnosis provides further supports for my position.

My answer is: a conventional attack could be morally justified in response to a cyberattack, but it is unlikely that an actual conventional attack could meet the justificatory conditions. I answer this because I assent to the following four propositions. First, the moral significance of harm to property is unaffected by how that harm is done. Suppose the Hoover Dam were destroyed. It could be blown by explosives. Perhaps it could also be caused to collapse by changing the way some system works. This system ensures that enough water is let through so as not to allow pressure to build up beyond the dam's capacity, and that system is vulnerable to remote interference—the sort of cyberattack meriting a Hollywood plotline. The harm of destroying the dam is the same in both cases, whether it is by explosives or cyberattack.

This is not to say that how the dam could be destroyed may not be morally salient in other ways. An explosion may result in direct loss of life and have required trespass for the rigging of the demolitions. A cyberattack might not have resulted in these wrongs. So some wrongs could be done by the first method of attack which would not be done by the second. Nonetheless, the harm to property caused by destroying the dam is the same.

Second, there is no international organisation that has both *de jure* and *de facto* authority to impose punishment on nation states. While it is arguable whether the United Nations has *de jure* authority, it unquestionably lacks *de facto* authority. In the absence of such an organisation, nation states are permitted to act on the basis of their own judgement about what it is just.

Third, war is justifiably seen as punishment and may be prosecuted as such by nation states given the lack of an authoritative international organisation to fulfil that function. While the idea that war may be a form of retributive punishment is admittedly unpopular at the present time there is a certain naturalness to it, as David Luban notes (2012). You raided our village, so we'll raid yours. A modern version: you bombed us, so we'll bomb you.

Given these three propositions, a nation may permissibly impose punishment on another nation by waging war when it has just cause. Suppose nation *B* has destroyed something valuable that belongs to nation *A* for no good reason. Nation *A* may permissibly destroy as punishment something of equivalent value belonging to nation *B*. Because the moral significance of harm to property is unaffected by how that harm is done, nation *A*'s punishment may permissibly take the form of a conventional strike in response to nation *B*'s cyberattack. An imagined scenario: China hacks into the Washington D.C. traffic control network, bringing the city to a standstill. In punitive response, the USA bombs the electricity sub-station that is a dedicated power supply for traffic lights in Beijing. A similar effect occurs, and repair costs in time and money are equivalent. The USAF has been extremely careful in their intelligence gathering and the operation is successfully conducted with no loss of life. I take the hypothetical response from the USA to be justified.

However, fourth, in the actual world rather than that of imagined scenarios, conventional attacks invariably lead to loss of life while cyberattacks seldom do. Such successful intelligence as that ascribed to the USAF above does not occur. So there will almost always be an asymmetry in the damage done by a conventional attack compared to a cyberattack. Loss of life is a weightier breach of rights than harm to property. It is a disproportionate punishment to retaliate against an enemy by killing people when they have solely harmed your property. Hence my answer: while it is *conceivable* that a conventional attack could be morally justified in response to a cyberattack, it is unlikely that an *actual* conventional attack could be morally justified.

This answer also provides a diagnosis of why the case is a hard one. For many will disagree with some or all of the first three of the propositions which I assent to. Disagreement with any will likely result in the conclusion that it is never morally permissible to respond to a cyberattack with a conventional attack. On this view, cyberattacks are not a legitimate *casus belli*.[1]

[1] Dipert concurs that cyberattacks may sometimes be a legitimate *casus belli*, but on the basis of the long-term value of conventional responses in minimising unprovoked cyberattack (2010, p. 400 ff.).

Because the ethical framework provided by property rights explains and justifies how cyberattacks are accommodated by Just War theory, it satisfies the second desideratum noted above (§ 2).

9.5 Justifying the Framework

The principal justification for my claim that property rights provide an appropriate framework for evaluating the ethics of cyberattacks is based on the claim that ICT systems are (invariably) owned by someone. Because ICTs are owned, their owners have property rights over their property not being interfered with by others.

If I was making an analogical claim—that the concept of property provided a useful analogy with which to think about the ethics of cyberattacks—then further justification would be required as to why the concept of property should be privileged for that role over other concepts which may be equally analogous. But I am making an identity claim: ICTs *are* a form of property. When I handed over the money for the PC now on my desk, the ownership relation changed. Previously the trader owned the computer, and after the money was handed over I then did. Computers and other information systems are owned by people and organisations. They therefore have rights which state what it is permissible for others to do in relation to those systems.

Noting that ICTs are a form of property does not entail that people do not hold other rights over them. I do not claim that breach of property rights explains *all* that is wrong with every kind of cyberattack. Some kinds of cyberattack are more akin to privacy violation. Other rights provide a basis for evaluating why these cyberattacks are wrong. Nonetheless, all cyberattacks involve a breach of property rights. So the framework applies to all cyberattacks. Furthermore, the value of my property and therefore the degree of harm done by breaching my property rights is affected by what else depends on my property not being interfered with.

A merit of using property rights as a framework is that it avoids objections to a sovereignty approach. Nation states have *de jure* sovereignty over territory. Violations of sovereignty are a central sufficient condition for *jus ad bellum* according to the Just War tradition, and thus a *casus belli*. Dipert rightly takes a sovereignty approach to be central to the Just War tradition, and rightly notes that cyberattacks do not 'fit' this approach. 'A cyberattack does not involve intrusions into the territory or airspace by soldiers or even by physical objects.... There is a nominal sense in which the photons [in Electronic Warfare] 'invade' the airspace of another nation, but that in itself seems harmless; foreign radio waves constantly pass through nations' airspaces without complaint' (2010, p. 397). Dipert has here anticipated James Cook's reply. Cook coins the neologism 'cyberation' to describe the 'goings-on' in cyberspace (Cook 2010, p. 414, 422 fn. 4). This is in analogy with aviation (air) and navigation (maritime). Armed with the analogy, Cook takes it as obvious that Just War theory applies to some cyberation but not all. While I am in sympathy with the broad point the analogy is unhelpful, and this for the reasons Dipert points out. The goings-on in cyberspace do not involve crossing of physical boundaries in the way that the goings-on at sea and in the air do. The analogy invoked by cyberation is a spatial one, but spatial boundaries are irrelevant in cyberspace. Indeed, 'cyberspace' is itself an unhelpful term, for it too invokes a spatial analogy (Stevens 2013).

In contrast, property can be harmed without a violation of sovereignty. So my approach sidesteps Dipert's objection to that approach. One need not reinstate borders to cyberspace to be justified in blaming those who conduct cyberattacks. One can do so because those who violate property rights are blameworthy. So long as violations of sovereignty are not required for embarking on a just war—and there is no reason why that should be—then Just War theory can accommodate cyberattacks.

Approaching the problem using the notion of rights has a further merit, for it avoids an objection to a simplistic form of applied ethics. This consists in the application of some normative theory to the domain in hand as follows. Suppose one is a rule-utilitarian. The principle *one ought to follow those rules which maximise greatest utility* then provides the following decision procedure. Compare the different rules which could govern action in the relevant applied domain. Judge what are the likely consequences of each being followed. Suppose the applied ethicist judges that Rule α would maximise greatest utility. Then the ethical imperative follows: one ought to follow Rule α. *Mutatis mutandis* for other normative ethical theories, whether they are different variants of consequentialism, or are deontological, or aretaic, and so on. The central principle(s) of each plausibly provide decision procedures which need only be applied to the domain at hand.

The problem with such a procedure is that disagreement at the normative level reiterates at the applied level. Failure to account for property rights is an objection to a candidate normative ethical theory. By using a notion that all plausible ethical views ought to agree on—viz. property—the applied framework I provide is not subject to pervasive disagreement at higher levels.

9.6 Objections

I conclude by replying to objections. Two are objections to the usefulness of my thesis that ICTs are property as a basis for moral judgement about the justice of cyberattacks. A third objection is against the contemporary viability of Just War theory, and *a fortiori* against the possibility of it accommodating cyberattacks.

a. *Harm to property is insufficiently serious*

Objection: Property is routinely harmed and owners' rights thereby breached. This is an everyday matter for the law to deal with, and not a cause for serious moral offense. Yet cyberattacks may be very serious indeed, causing widespread disruption of a decidedly non-everyday sort. In scenarios which one day may be actual, cyberattacks are appropriately responded to by airstrikes. Harm to property is insufficiently serious a classification of the harm done by such cyberattacks, and insufficiently serious a justification for this kind of response.[2]

Reply: While much harm to property is routine, some is not. It depends on how valuable that property is. Causing a dam to collapse is extremely serious. Dams can be extremely valuable in both financial and practical terms—take the Hoover Dam as an example—so it is an extremely serious matter to cause harm to them. *Mutatis mutandis* for those ICTs the proper function of which is required for the delivery of essential services like water and electricity.

b. *Rights-talk is controversial*

Objection: Rights-talk is controversial. It is just as controversial as any of the classical ethical theories. It is no merit of my approach that it avoids pervasive disagreement.

Reply: I use a thin notion of rights which is committed to only the following two propositions. First, there are ways in which a person ought to be treated. Two, ownership of property changes the ways in which others may treat the owner and that property. Section 3 above explains what some of these ways are.

Moral error theorists like J. L. Mackie disagree with the first proposition: there is no such thing as obligation or desert (Mackie 1977). Certain kinds of anarchist disagree with the second. But beyond these fringe positions the above two propositions command widespread acceptance.

c. *Just War theory's 'rickety meta-ethical foundations'*

Objection: Just War theory relies on 'rickety meta-ethical foundations'. So it ought to be rejected. Dipert notes that the tradition originates historically out of natural law theory via Thomas Aquinas and Hugo Grotius. 'Yet few philosophers have a metaphysics that would allow them to maintain such a view.... [Thus t]he only mete-ethical route left to many theorists is one based upon "intuitions", legitimized by the "reflective equilibrium" of John Rawls' (Dipert 2010, pp. 393–394).

Fill out the suppressed premises of his argument as follows. Not only is there an historical relation between Just War theory and the natural law tradition, there is a justificatory one. If one is justified in endorsing the central claims of the natural law tradition, then one is justified in endorsing those of Just War theory. But it is controversial whether one is justified in endorsing the natural law tradition. Those that do not are not justified in endorsing Just War theory.

Reply: The argument is fallacious. Dipert has denied the antecedent. Grant that one is justified in endorsing Just War theory if one is justified in endorsing the central claims of the natural law tradition. Nonetheless, one may deny the central claims of the natural law tradition while endorsing those of the Just War theory, just if one takes that theory to be justified on another basis. Denying the validity of natural law theory has no logical relation to the possibility of another justification of the Just War conditions.

[2] I am grateful to Luciano Floridi for pressing this.

What alternative justifications are there for Just War theory? I need not take a stand. I have used the notion of rights, and a number of different ethical theories may plausibly claim to justify rights.

Rejoinder: My reconstruction of Dipert's argument is uncharitable. The conditionals asserting justificatory relations I ought to have ascribed to him give necessity conditions as well as sufficiency ones. One is justified in endorsing the central claims of Just War theory if *and only if* one is justified in endorsing those of the natural law tradition. Then Dipert would not have denied the antecedent.

Reply to the rejoinder: Granting the reconstruction, his argument is then unsound. For it is false that one is justified in endorsing Just War theory only if one endorses the central claims of the natural law tradition. This is for the same reason as above, that rights are sufficient to ground the theory.[3]

References

Arquilla, John. 1999. Can information warfare ever be just? *Ethics and Information Technology* 1:203–212.

Arquilla, John. 2010. Conflict, security and computer ethics. In *The Cambridge handbook of information and computer ethics*, ed. Luciano Floridi, 133–148. Cambridge: Cambridge University Press.

BBC. 2007. Israel admits air strike on Syria. BBC news. http://news.bbc.co.uk/1/hi/7024287.stm. Accessed 21 Apr 2012.

BBC. 2012a. Man who hacked into abortion provider website jailed. BBC news. http://www.bbc.co.uk/news/uk-17706621. Accessed 21 Apr 2012.

BBC. 2012b. Two arrests over Scotland yard terror line hack. BBC news. http:bbc.co.uk/news/uk-17698528. Accessed 17 Apr 2012.

Betz, David J., and Tim Stevens. 2011. *Cyberspace and the state: Towards a strategy for cyberpower*. London: Routledge.

Coady, C. A. J. 2008. *Morality and political violence*. Cambridge: Cambridge University Press.

Cook, James. 2010. 'Cyberation' and just war doctrine: A response to Randall Dipert. *Journal of Military Ethics* 9:411–423.

Denning, Dorothy E. 2008. The ethics of cyber conflict. In *The handbook of information and computer ethics*, ed. Kenneth Einar Himma and Herman T. Tavani, 407–428. Hoboken: Wiley.

Dipert, Randall R. 2010. The ethics of cyberwarfare. *Journal of Military Ethics* 9:384–410.

Falliere, Nicolas, Liam O. Murchu, and Eric Chien. 2011. W32.Stuxnet Dossier, version 1.4. Symantec. http://www.symantec.com/content/en/us/enterprise/media/security_response/whitepapers/w32_stuxnet_dossier.pdf. Accessed 21 Apr 2012.

Furness, Hannah. 2012. Team poison: The interview. The Telegraph. http://www.telegraph.co.uk/news/uknews/crime/9200813/Team-Poison-the-interview.html. Accessed 17 Apr 2012.

Gasparre, Richard E. 2008. The Israeli 'E-tack' on Syria—Part II. Air force technology. http://www.airforce-technology.com/features/feature1669/. Accessed 21 Apr 2012.

Guthrie, C., and M. Quinlan. 2007. *The just war tradition: Ethics in modern warfare*. London: Bloomsbury.

---

[3] I am grateful to Mariarosaria Taddeo and Luciano Floridi for organising and inviting me to present at the workshop which was the spur for this volume, and to attendees for their comments and questions on the paper. My thanks to Alex Oliver, Randall Dipert and Tim Stevens for criticism of an earlier draft.

Halperin, D., T. Heydt-Benjamin, B. Ransford, S. Clark, B. Defend, W. Morgan, K. Fu, T. Kohno, and W. Maisel. 2008. Pacemakers and implantable cardiac defibrillators: Software radio attacks and zero-power defenses. *IEEE Symposium on Security and Privacy* 2008:129–142. http://ieeexplore.ieee.org/xpl/articleDetails.jsp?arnumber=4531149&sortType%3Dasc_p_Sequence%26filter%3DAND%28p_IS_Number%3A4531132%29

IAEA. 2011. Implementation of the NPT safeguards agreement in the Syrian Arab Republic. http://www.iaea.org/Publications/Documents/Board/2011/gov2011-30.pdf. Accessed 21 Apr 2012.

Leyden, John. 2007. Israel suspected of 'hacking' Syrian air defences. The Register. http://www.theregister.co.uk/2007/10/04/radar_hack_raid. Accessed 21 Apr 2012.

Luban, David. 2012. War as punishment. *Philosophy and Public Affairs* 39:299–330.

Mackie, J. L. 1977. *Ethics: Inventing right and wrong*. Harmondsworth: Penguin.

Melman, Yossi. 2011. Outgoing Mossad chief: Iran won't have nuclear capability before 2015. Haaretz. http://www.haaretz.com/print-edition/news/outgoing-mossad-chief-iran-won-t-have-nuclear-capability-before-2015-1.335656. Accessed 21 Apr 2012.

Rid, Thomas. 2011. Cyber war will not take place. *Journal of Strategic Studies* 35:5–32.

Ross, W. D. 1930. *The right and the good*. Oxford: Clarendon Press.

Rowe, Neil C. 2007. War crimes from cyberweapons. *Journal of Information Warfare* 6:15–25.

Rowe, Neil C. 2008. Ethics of cyber war attacks. In *Cyber warfare and cyber terrorism*, ed. Lech J. Janczewski and Andrew M. Colarik, 105–111. London: IGI Global.

Rowe, Neil C. 2010. The ethics of cyberweapons in warfare. *International Journal of Technoethics* 1:20–31.

Stevens, Tim. 2013. Information warfare: A response to Taddeo. *Philosophy and Technology* 26:221–225.

Taddeo, Mariarosaria. 2012. Information warfare: A philosophical perspective. *Philosophy and Technology* 25:105–120.

Tavani, Herman T. 2002. The uniqueness debate in computer ethics: What exactly is at issue, and why does it matter? *Ethics and Information Technology* 4:37–54.

Tavani, Herman T. 2005. The impact of the internet on our moral condition: do we need a new framework of ethics? In *The impact of the internet on our moral lives*, ed. Robert Cavalier, 215–237. Albany: SUNY Press.

Chapter 10
Virtue in Cyberconflict

Don Howard

Abstract Both offensive and defensive cyber operations require decision making on short time scales and often in a partial vacuum of explicit norms for cyberconflict. It is argued that, under these circumstances, the virtue ethics approach to ethical decision making offers significant advantages over rule-based ethics in the training of cyber warriors. Cultivating the moral integrity of the cyber warrior should be the goal, moral integrity now understood as moral character in the form of settled habits or dispositions to act appropriately in different contexts. In shaping the character of cyber warriors, attention should be paid to both moral and intellectual virtues. Explicit training regimens are sketched and applications to model scenarios are explored. As cyberconflict becomes more automated, with the morality moving into the machine, the advantages of the virtue ethics approach will become even more pronounced.

> *Socrates: But, oh heavens! shall we condescend to legislate on any of these particulars?*
> *Adeimantus: I think, he said, that there is no need to impose laws about them on good men.*
>
> Plato, The Republic, Book IV

10.1 Introduction: Is Integrity a Problem or a Resource?

"Warfare in a New Domain: The Ethics of Military Cyber Operations" was the topic of the 2012 McCain Conference at the U.S. Naval Academy[1]. It was a signal event, drawing a distinguished group of participants from the military, the intel-

[1] A video record of the 2012 McCain Conference is available here on the website of the US Naval Academy's Stockdale Center for Ethical Leadership, which sponsored the conference: http://www.usna.edu/ethics/publications/mccain2012.php.

D. Howard (✉)
Department of Philosophy and Reilly Center for Science, Technology, and Values,
University of Notre Dame, Notre Dame, USA
e-mail: dhoward1@nd.edu

L. Floridi, M. Taddeo (eds.), *The Ethics of Information Warfare*,
Law, Governance and Technology Series 14, DOI 10.1007/978-3-319-04135-3_10,

ligence community, the executive office, government agencies, NGOs, academia, and private business. Following in the wake of the 2009 National Research Council's study, *Technology, Policy, Law, and Ethics Regarding U.S. Acquisition and Use of Cyberattack Capabilities* (National Research Council 2009), the 2012 McCain conference marked the high point, to date, in public engagement with the ethics of what is rapidly emerging as the most important new technology and arena of military conflict.

Among the most interesting moments in the conference occurred during the panel on "Operational Perspectives on Cyber Operations." Air Force Lt. Col. Matteo Martemucci, National Security Affairs Fellow at the Hoover Institution for 2011–2012, is the former commander of the Air Force's 315th Network Warfare Squadron, the Air Force's computer network attack unit, which reported to US Cybercommand and conducted computer network exploitation and computer network attack. Col. Martemucci said that most of the work of the 315th was, from an ethical point of view, "clean and good," and that, so far, it had been mostly "small scale." But one concern was that the partial vacuum of law in which offensive cyber operations are conducted and the speed at which decisions must be made leaves us reliant upon the "integrity" of the operator to a greater extent than with many other technologies and in many other conflict arenas. Col. Martemucci was later pressed on this point by myself and by Maj. Gen. Charles J. Dunlap, USAF Ret., Executive Director of the Center on Law, Ethics, and National Security, at the Duke University Law School. The question was, simply, whether our being forced to rely on the "integrity" of the operator is a good thing or a bad thing.[2]

One does not want to overstate the novelty of the issue in cyberconflict. Combatants often find themselves having to make snap decisions under conditions of imperfect ethical and legal guidance, and other, new, warfighting technologies likewise press operators in ways less commonly encountered with bullets, bayonets, and bombs. Nonetheless, there is at least a difference of degree, perhaps rising to a difference in kind, with cyberconflict. This difference in degree or kind is a consequence of both the speed with which information moves in cyberspace and the complexity of the networks within which and through which cyberconflict is waged. Ethical decision making in such a setting is often not a leisurely affair. In the time it takes to consult a Judge Advocate about the legitimacy of a target, such as a hijacked private corporation's server in a neutral nation, a brief attack window might well have closed.

Maximizing the likelihood of ethical decision making under such circumstances is a challenge. Even if the International Law of Armed Conflict were to catch up with the new technology of cyberconflict,[3] and even if the law were so well written as to render determinate every potential decision, which is, of course, impossible, the complementary problems of speed and complexity would remain. Hence the sense of commanders like Col. Martemucci that we have no choice but to rely upon the "integrity" of our cybercombatants. So, again, is this a good thing, or a bad thing?

[2] There is, as yet, no transcript of conference presentations nor is any publication planned. But see the videos of the individual talks and panels as cited in note 1.

[3] The US government has recently restated its position that the extant international law of armed conflict governs cyberconflict. See Koh 2012.

10.2 Virtue and Ethical Decision Making in General

Begin with the question of what it means to rely on a cybercombatant's integrity as opposed to alternative models of ethical decision making. I assume that the contrast intended is with decision making under rules, a mode of ethical decision making often denoted "deontological" and associated, paradigmatically, with the moral philosophy of Immanuel Kant.[4]

A rule asserts a general prohibition or obligation. "Perfidy is never permitted in war." "Harm to non-combatants is not permitted unless an unavoidable consequence of a legitimate attack against a legitimate target." "One must always identify oneself by uniform or insignia as a member of a properly-constituted fighting force." One then decides the rightness or wrongness of an act by first deciding whether the act falls within the scope of the relevant rules. Is a Trojan-horse attack an act of perfidy? If so, it is not permitted. Is taking down a hijacked server in a neutral country necessary for defense against a potentially crippling denial of service attack on a nation's financial infrastructure? If so, it may be permitted even if taking down that server causes financial or other harm to innocent citizens of the neutral nation in which it is located. Ethical decision making so modeled is what is presumed in the body of international law and ethics of warfare.

Integrity, on the other hand, is a virtue, or, perhaps, a cluster of virtues. The virtue-based or aretaic model of ethical decision making looks quite different.[5] Indeed, it can be misleading even to speak of "decision making' within this framework. A virtue like courage, generosity, or truth-telling is a habit or a settled disposition to act. It differs from a vice in being directed toward a good. Acting in accordance with a virtue is not a matter of recruiting the will to move action as indicated by the conclusion of a practical syllogism. Acting in accordance with a virtue is, simply, acting as disposed to act or doing what the virtuous habit inclines one to do. As a settled disposition, it need not involve any logical epicycles at all. One is virtuous simply by virtue of tending, in the main, to do what virtue asks of us.

If there is anything like decision making in this domain, it concerns not the subsuming of a special case under a rule, but the assimilating of a present situation to relevantly similar situations in which like circumstances elicited like action. It would be a mistake to intellectualize such assimilating beyond a certain point. Moral phenomenology is helpful here. The virtuous individual does not really reflect upon the similarity in question. If asked to describe the experience, the virtuous individual reports just recognizing the situation as one in which courage was appropriate and so acting courageously. As virtue is becoming settled habit, as in

[4] Kant's *Groundwork of the Metaphysics of Morals* (1785) is the standard source. According to Kant, only the good will is good in an unqualified way, and only the person who acts from pure duty is the person of good will. Acting from pure duty means acting in accordance with a rule or principle, the highest of which is the Categorical Imperative: "I ought never to act except in such a way that my maxim should become a universal law." A helpful recent introduction to ethical theory is Shafer-Landau 2011.

[5] Aristotle's *Nicomachean Ethics* is generally regarded as providing the most highly articulated, early statement of the program of virtue ethics. Shafer-Landau 2011 is, here too, a good introduction.

the young or those otherwise new to virtues of a certain kind, more explicit reflection might be required, sometimes after the fashion of asking with the person of acknowledged virtue, the moral exemplar, would do in similar circumstances. But mature virtue is almost automatic in the way in which it feels to the moral agent. I told the truth not because I paused and deliberated the rightness of truth telling in this case (though, of course, I might do that in highly problematic situations), and decided upon honesty. No, I told the truth because I am a truth teller. That's all there is to it. How one achieves such a state of moral maturity is the question of the cultivation of virtue, to which I turn below.

For now the point to emphasize and prize is that, within the aretaic setting, the path from virtue to action is a short one, precisely because a virtue is a disposition to act. I like the metaphor of the moral athlete. When learning a new sport, one often has to pause and think. Should I take four or five steps between two hurdles? Do I begin to open from the tuck position at the top of the arch or as my dive is approaching the downward vertical? But such thinking is fatal to polished performance. The gold medal winner is the athlete in whom the entire race or dive is more or less automatic. Musical performance affords another set of analogies. The beginner has to count out the beat and think about where to find the G#. The master, like Arthur Rubinstein or Mitsuko Uchida, just plays flawlessly. We invent words like "Fingerspitzengefühl" to describe such competence.[6]

10.3 Virtue and Ethical Decision Making in War

As in life more generally, so too in war.[7] Prominent among the virtues discussed by Plato, Aristotle, and the other ancients from whom we learned the language of virtue was courage. While not exclusively a martial virtue, its central role in Greek thinking about the moral life reflects the common role of members of the Greek polis as citizen-soldiers. Courage teaches us much about virtue. For Aristotle, each virtue is a mean between two extremes. Excess of courage is rashness; defect of courage is timidity. Where the mean lies can vary from person to person and situation to situation, and knowing how to fine tune the level of one's courage to the situation

[6] Contemporary ethical theory normally identifies consequentialism as a third general framework for ethical decision making, this distinguished from deontology and aretaic ethics by its assertion that an act qualifies as moral or immoral on the basis of its net consequences for human happiness or suffering. The utilitarianism of Jeremy Bentham and John Stuart Mill is the best known variety of consequentialism. Yet again, see Shafer-Landau 2011 for the fundamentals. For the purposes of this essay, I shall pass over the consequentialist alternative to deontology and virtue ethics, because, if speed and facility in ethical decision making are the issue, then the problem is hardly helped by the proposal to carry out a tedious and complicated exercise in what is sometimes known or derided as the "hedonic calculus," however much, in fact, we often reason or think we reason in consequentialist terms.

[7] Shannon Vallor discusses the concept of military virtue in connection with armed robots in her contribution to this volume. See Vallor 2013.

is a mark of the difference between the morally mature and the moral obsessive, who rigidly adheres to a rule heedless of circumstance. The trainee might waver, paralyzed by indecision or fear. But the courageous veteran acts surely and in the appropriate measure. So doing is the measure of courage as virtue.

Other virtues characterize war fighters then and now. Honor, loyalty, and pride are often discussed, integrity sometimes taken to subsume them all and sometimes styled, instead, a kind of meta-virtue bespeaking an overall seriousness of moral purpose and responsibility. Especially in the training of the officer corps, great stress is still laid upon the cultivation of such virtues, and their possession is what sets the military officer apart in the minds of many.

Outside of the military, virtue as the concept around which to frame our moral discourse has not had the happiest of careers of late. The popularity of virtue ethics in the academy since the publication of Alasdair MacIntyre's *After Virtue* (MacIntyre 1981) contrasts with and is, perhaps, driven by academic concern about the uncritical consequentialism so typical of contemporary debate about issues of small and large moment. This is not the place to analyze such trends in detail, so I settle only with a remark about the ubiquity of risk and cost-benefit analysis as the ubiquitous tools of public policy making.[8]

More to the point of this essay is the career of virtue in the military. Space again permits no analysis beyond anecdote. But my own experience of teaching ROTC cadets and the testimony of friends and colleagues within the military point up the greater difficulty one has today in eliciting sympathy for notions like honor and integrity among younger officers in training. One reason for this state of affairs is surely that our military finds itself ever more in conflict situations where we confront enemies who seem to lack virtue or whose virtues are not easily recognized as such because of cultural difference or extreme differences of circumstance. A Wehrmacht officer might be recognizable as an honorable individual, however great the carnage of World War II, likewise a Japanese officer true to the Bushido Code. But the Vietnam War put greater pressure on our ability to see our enemy as like in kind to us, and now, even more so, the kinds of conflict we experience in places like Somalia and Afghanistan. Virtue thrives in the company of virtue, even when the goods served might differ. Virtue struggles to survive in a vicious world or a world perceived as vicious.

I think that I see, however, a different source of pressure on the place of virtue in the military, one more relevant to the theme of virtue in cyberconflict. It is the very technologizing of conflict that has us worrying about the ethics of cyberconflict in the first place. This is an old trope. Our way of regarding the martial virtues grew up, as did our law of armed conflict, in a world wherein war and combat looked very different than they do today. War was state-on-state conflict. Combat was man-to-man. About this latter we have at least secure intuitions of what courage and honor

---

[8] An alternative approach to policy making drawing partly on the virtue ethics tradition starts from consideration of a policy's impact on the development and maintenance of human capabilities. See Nussbaum 1988 and 2000 and Sen 1985 and 2009, along with Nussbaum and Sen 1993 for representative and exemplary discussions.

and loyalty mean. Then came new machines of war. Already at the Battle of Crécy in 1346, the French thought that the English had shown no honor by employing the long bow with massed archers. But certainly by the time of the strategic bombing campaigns of World War II, worry set in that, as the technology of war depersonalized war, the pilot and bombardier never seeing their victims thousands of feet below, it left ever less room for virtue as most relevant distinction of the capable war fighter. Fortitude in the face of a flak attack and loyalty to one's fellow crew members still mattered. But it mattered more that one remembered and could use one's high-school trigonometry and vector analysis to place the bomb on the target or the machine-gun fire on the enemy fighter.

Fast forward to the world of drones and smart munitions. The operator putting munitions on a target now sits thousands of miles away in a trailer, out of the way of immediate retaliation. This mode of combat takes its toll, to be sure. But the challenge of staying alert throughout an 8 h shift is hardly comparable to that of remaining brave when you've just seen your best friend eviscerated by shrapnel, you haven't slept in 48 h or eaten in the last twelve, and you know that a superior force is about to assault through the tree line fifty yards in front of you. When the bridge comes to resemble an IT control room and the headquarters battalion a cadre of systems engineers, one might not be faulted for thinking that calculation has displaced courage.

10.4 The Novel Challenge of Virtue in Cyberconflict

If the technologizing of war is part of what renders honor and courage irrelevant, how, then, can it be that relying upon the integrity of the operator is a way to insure ethical action in cyberconflict? Isn't cyberconflict the ultimate in technologized warfighting? Isn't this an arena in which algorithms should replace virtues?

To understand why one might think otherwise, why one might think that a turn to virtue is what is needed in cyberconfclict perhaps more than in any other conflict arena, let's first think a bit more about the scope of the virtues. So far we have concentrated, as does most discussion of virtue in the military life, on what Aristotle categorized as *moral* virtues. But foremost among the virtues for the Greeks was wisdom, an *intellectual* virtue, possession of which was more or less a necessary condition for possession of all the rest. And Aristotle further distinguished a total of five principal and four subsidiary intellectual virtues.

First among the principal intellectual virtues are three virtues specific to theoretical knowledge: *Sophia* or wisdom in this more specific setting, *Episteme*, which means roughly scientific or empirical knowledge, and *Nous* or reason in the sense of rational or logical thinking. *Phronesis* is practical wisdom, the variety of wisdom specific to the moral life. *Techne* names the kind of knowledge found in the artisan, what one might term "knowing how." The subsidiary intellectual virtues have a role to play throughout and include such as *Euboulia*, the habit careful and proper deliberation leading to good ends, and *Deinotes*, a kind of cleverness in achieving ends of any kind.

Renewed scholarly interest in virtue ethics has led to a similar renewal of interest in intellectual virtue as a way of reframing many classic questions in epistemology or, rather, moving the center of epistemological inquiry to new kinds of questions.[9] There is much of interest in this domain, as with recasting questions about the truth of theories in terms of the truthfulness and, more generally, the intellectual integrity of the theorist. But our topic is, specifically, virtue, now including intellectual virtue, in cyberconflict.

A mere inspection of Aristotle's list of intellectual virtues makes clear their relevance to cyberconflict. The cyberwarrior requires scientific knowledge, reasoning ability, the skill of the digital craftsperson, and cleverness, along with moral wisdom. So, too, have warriors throughout history—one thinks of the Gen. Patton who knew the history and theory of war from Thucydides to Clausewitz or the GI who could jury-rig just about anything—but war fighting in an ever more technical age puts still more of a premium upon these distinctively intellectual virtues.

Two questions now intrude. The first is whether anything of importance turns upon our restyling scientific, empirical knowledge, say, as a virtue rather than just the body of theory and fact that we ordinarily take it to be. Indeed, is it not odd and misleading to call it virtue? Scientific knowledge is best understood as codified in a set of propositions or models. Where is the virtue in Einstein's gravitational field equations? Let us not wander into unhelpful arguments about a fact/value distinction, however relevant those issues might be in a more extensive inquiry into the issue at hand. Keep the focus on the moral formation of the cyberwarrior. Aristotle (or at least an Aristotle who lived to see the linguistic turn in twentieth century philosophy; allow this innocent anachronism) would not deny that scientific knowledge can be represented propositionally. But Aristotle stands in a long tradition of thinkers up to and beyond John Dewey in the twentieth century,[10] all of whom think it important to put philosophical attention more on the knowing than the knowledge, to emphasize the capacities possessed by the knower as an epistemic agent, the point being that what one knows is less important than what one can do and does with that knowledge. Empirical science, from this perspective, is a way of knowing.

The second question is, what happened to the moral virtues? We were thinking, to begin with, about the ethics of cyberconflict. I think that we do well to follow Aristotle in regarding both the intellectual and moral virtues as essential parts of character formation. It is not just that Aristotle designated wisdom as the highest virtue, and that he believed, with Plato, that to know the good was to do the good.[11] Aristotle understood the moral virtues as contributing to the intellectual life and the intellectual virtues as contributing to the moral life. There is, for example, such a thing as intellectual courage in pursuing a line of investigation in the face of strong

[9] For a helpful introduction to the contemporary discussion of intellectual virtue, see DePaul and Zagzebski 2007.

[10] Dewey's "adverbial" theory of knowledge is given a definitive statement in Dewey 1929.

[11] "To know the good is to do the good" is a capsule formulation of the Aristotelian idea of Eudaimonia, which finds expression in both the *Nicomachean Ethics* and the *Eudemian Ethics*. More subtly expressed, it is the idea expressed above in the text about the intimate relationship between the moral and intellectual virtues.

opposition from one's peers or into unexplored terrain. And, among those lacking in wisdom and discernment, munificence or liberality can easily become a foolish tendency to say "yes" to every huckster peddling heart-wrenching propaganda on behalf of a fraudulent charity. For Aristotle, the good person combines intellect and morality in one balanced whole.

10.5 Cultivating Both Moral and Intellectual Virtue in the Cyberwarrior

The discussion so far has been highly theoretical. Let's bring back down to earth, as it were, by asking how the perspective so far sketched might make a practical difference in the training and evaluation of cyberwarriors. How do we train people for virtue?

Recall, first, that virtues are habits of settled action directed toward a good. Our question, then, is how we inculcate the desired habits. The answer is two-fold: practice and modeling. Start with practice. A habit is a disposition, not mechanical, repetitive behavior. We want to produce not Pavlovian salivators, but moral agents capable of acting surely in a measured way in a wide variety of settings. As with the learning of any skill, it starts with drill. The piano student spends many hours turning scales into mechanical routine. But drill is only the starting point for pianist and the moral novice. Drill produces the basic repertoire of actions around which more sophisticated habit grows. Once one learns those basics, then practice is the path to excellence. Only practice teaches the aspiring gymnast how to respond to every possible variation of pressure, angle, and impact when landing on the balance beam. Only practice teaches the young quarterback how to recognize and respond to every different defensive set. There is a fine structure to this evolving competence, one relevant to the next point to be made about the inculcation of moral virtues. Modeling is the key. One first learns the moves appropriate to a given situation, whether a three-note trill or a down-and-out. And then another situation, and yet another. When next one is confronted with a new situation, the response of one growing in skill or virtue is to see it as like a situation for which the appropriate action is already known and to act accordingly, trying this or that slight variation. As the repertoire of model situations grows, so grows the competence of the athlete, the performer, or the moral agent.

Modeling is key in expanding the range and sureness of habits that are more than merely mechanical. It is key, as well, in the manner of instruction we offer. What is the role of the teacher? Rote instruction in basic rules and principles is as necessary a starting point as is drill. But that kind of training carries the learner only so far. Excellence in sports, performance, or morality is best produced by the teacher who functions as an exemplar. We often speak of moral exemplars like Jesus, Ghandi, or Audie Murphy without always appreciating the power of such examples in shaping behavior. We speak glibly about teenagers "modeling" on their peers, idols, or authority figures, without always understanding that "do as I say, not as I do" is precisely the opposite of the formula that achieves maximum effect. Professional educators often overlook the fact that students spend more time emulating their manner than memorizing for the next test.

That modeling is essential to developing skills and habits in other domains is almost a commonplace. The coach takes the ball from the basketball player's hand and says, "Watch, this is how one does a fade-away jumper." The vocal coach places the soprano's hand on her own abdomen and says, "Feel what I'm doing; see, you breathe from here." The blacksmith says to the apprentice, "Listen, hear the sound the iron makes when its temperature is right and you strike it at the right angle; now you try it." And sometimes the point is appreciated in the classroom. The prize-winning teacher says, "Let me show you how I would do problem number 1, then you try problem number 2."

That modeling is equally essential in character formation is less of a commonplace. But Plato understood it. That is why, in Book III of *The Republic*, he banned the form of poetry that we would call drama in the ideal state. It was because he understood that, even were the exhibited behavior expressly condemned in a play, the very exhibition could overwhelm the damning words in encouraging imitation. Another place where this point is also sometimes well understood is the military. Think of the many stories we tell about and the praise we offer to commanders who lead by example.

Virtue both moral and intellectual starts in drill, grows with practice, and is nurtured by example. Is there guidance to be found here in how to train cyberwarriors? I think that there is, and this especially as it pertains to the goal of maximizing the likelihood of ethical action in cyberconflict. Let me first confess, however, that I do not know exactly how we currently train operators in cyberconflict, nor would I want a civilian like myself to have access to anything more than the most basic training regimen. Let me, therefore, present my suggestions as a hope that we do train in something like the following manner and, otherwise, a gentle counsel to consider the option.

The training should begin with drill, grow with practice, and be nurtured by example. Only those with all the basic skills need apply. Start, then, with rote learning and memorization of the fundamental advanced skills essential in the tool kit of any operator. But then devote the bulk of the training to practice guided by the example of the veteran. We know how to do this and we have lifetimes of experience in nearby fields. I think especially of the way we train pilots, astronauts, and law-enforcement professionals. First they have to learn the basic principles and techniques of flight, but then pilots and astronauts spend endless hours in simulators. Sometimes they practice routine landings and takeoffs or launches. But they are also presented with a wide variety of scenarios that hone and test their responses to every kind of novelty, contingency, and emergency. And they have to be recertified from time to time with similar testing on the simulators.

Of course what's lacking in flight simulators is the moral dimension of flight. Or is it? Do we present scenarios where the pilot has to complicate the decision making by weighing risk to passengers against risk to people on the ground. Do I veer left at the last minute to avoid that small settlement and risk death to passengers and crew, or do I guess that at mid-day so few people will be at home as to warrant my landing the plane on Main Street? One place where such moral simulations are standard is in the training of law enforcement officers, as with an active firing range, where, along with having to shoot straight one must decide whether a suddenly appearing figure is a person with an infant or with a gun. My impression is that comparable scenarios are part of the training of some combat troops, especially in training for urban warfare and in the training of special operations forces.

Experience suggests that practice makes perfect in the moral simulator as well as the flight simulator. One real human life rarely confronts the individual with a truly wide array of moral challenges, and most of our moral puzzles are mundane. Do I keep that $ 20 bill I found at the front door of my office this morning, or do I make the effort to find the owner, time consuming though that might be? Do I run that red light when no other motorists are in view in order to make my next appointment? A virtual moral simulator can do so much more. For many of us, this role is played by fiction. But even there the dilemmas are often predictable and cartoonish, and as often as not we're secretly identifying with the bad guy, not the good one. For those of us more likely to find ourselves in moral situations of real moment, time spent in deliberately constructed, role playing and simulation is probably time well spent. If cops and SEAL teams can learn better to discriminate between armed criminals and bystanders or terrorists and taxi drivers, why is a comparable gain not to be expected in the moral performance of the cyberwarrior?

War gaming is, of course, a common exercise in the military. But this is usually only for the benefit of large units, not individual operators, and the focus is mainly on strategy and tactics, not the morality or legality of either large plans or individual acts. We set the law of armed conflict as a constraint, but rarely is moral choice the point of the exercise or even the hidden agenda.

What I propose, again, is that simulation with the ethics of cyberconflict as one of the explicit training goals be made a major part of the cyberwarrior's training and recertification. What is the gain? More reliably moral action is one expected gain. But to recur to the problem with which we started—the need for rapid action in a complex environment—gain along such lines is to be expected as well. Think again of the astronaut, piloting a shuttle and confronted now with a sudden loss of control for a crucial control surface during a critical phase of descent through the upper atmosphere. Time does not permit a first principles derivation of system response to ten different possible actions—maneuvering thrusters, attitude change, altered angle of descent, etc.—for regaining control of the craft. One must act and act quickly. The pilot with the best chance is the one who feels in his or her bones what is happening, knows, as if instinctively, the consequences of each possible reconfiguration, and so can recover without thinking, as it were. How to develop such ability? Practice, practice, practice. Think now of the cyberwarrior, peeking inside the server controlling the deposits and transactions of a major Gulf state bank. We have discovered the hidden assets of a major terrorist organization whose activities can be crippled with one more keystroke, an opportunity perhaps never to be had again. Then it is noticed, with just ten seconds to go before the server's intrusion detection software closes the back door through which entrance was gained, that, cleverly co-mingled with those assets, are heretofore concealed, off-shore holdings of a dozen major US corporations whose losses from that keystroke would be in the billions of dollars. Why are those assets there? Who is responsible for the co-mingling? The concealment, in any case, is illegal, is it not? But the damage to the US economy could be severe. Were they really US assets in the first place, or just disguised to appear so? Perhaps it's just a clever ruse, an extra layer of protection of those terrorist assets against just such a cyber attack. How does one weigh the presumptive property rights of the corporations against the right to life of the terrorist organization's potential targets? What does justice require? What does fairness demand? What will my commanding officer say? What does prudence require? What does courage command? What is the right thing to do? Call in the Judge Advocate, and the opportunity is lost forever. Hesitate, and the opportunity is gone. Err on the side of caution? Strike and risk economic disaster as the cost of saving, potentially, thousands of lives?

The example is contrived and extreme, to be sure. But it is not so otherworldly. One remembers reports of a missed opportunity to take out bin Laden in the mid-2000s when chain of command delays and uncertainty about identification and collateral damage meant inaction. But no one is prepared for such moments by our mundane moral experience. And those who have grown through real world experience of enough such situations are too rare to staff or even supervise all the many stations where such judgments are made. Training by simulation, moral simulation, can supply the lack.

Apprenticeship, imitation, and learning by human exemplar was the other form of habit formation by modeling discussed above. If this, too, is a way to grow the sought-for competence in moral and intellectual virtue among cyberwarriors, then there are further implications for training. Perhaps apprenticeship should become the way to train cyber operators. The rookie is paired with a veteran, just as pilot and co-pilot share a cockpit, the co-pilot learning by example and being entrusted with a gradually expanding array of tasks until he or she is good enough to shift into the pilot's seat. The apprenticeship model might well encounter resistance from veteran operators and a larger institutional culture that favors rapid promotion into new responsibilities and variety of experience in grade over extended service in a single assignment. But some highly successful warrior cultures—one thinks of the Plains people of the Americas—long trained their young in the arts of war in very much this fashion, and Plato explicitly commends it in *The Republic*.

10.6 Conclusion: What Happens When the Morality Is in the Machine?

The moral predicament of the contemporary cyberwarrior could well be a short-lived one. The iconic image of cyber conflict in the early twenty-first century features row upon row of Chinese or North Korean personnel, each at an individual terminal with his or her own target list and tasking. But it is surely on a matter of time before much of the human role in cyber operations is taken over by the computer itself. If autonomous drones and robots are just over the horizon, more autonomy in cyber operations is to be expected as well. The cost equation will drive this development as it already drives the replacement of expensive kinetic deterrence by cheaper cyber deterrence. Human operators are expensive to train and maintain. Much of the dreary routine work of probing and testing, decipherment and password breaking is surely already automated. If soon we will trust the facial recognition software to pick out the terrorist in the crowd and the drone then to make the kill shot, with at best a human on the loop, not in the loop, why not also trust the computer code to find the target server and take it out, perhaps even covering its own tracks as it leaves. It should take only a moment to recognize in this a sketch of the Stuxnet concept. The future is now.

Whither, then, the morality? I discussed above the way in which the increasing technologizing of war puts pressure on the notion of virtue in war. As cyber conflict becomes ever more algorithmic, is there a place for morality as virtue or morality under any description? Advocates of autonomy in weapons systems argue that autonomous systems can, in principle, be more moral than human operators. Not only can the computer run through a decision tree far more quickly than a human, the computer feels no pain, no fatigue, no anger or anguish. Program in the law of armed conflict and trust the machine to a better job than any human ever did of distinguishing friend from foe and combatant from non-combatant, calculating the permissible of casualties allowable as collateral damage from a legitimate targeting, executing the strike, and doing the post-operation outcomes analysis, including the legal and ethical outcomes.

I am not one to object to such visions of future combat on the basis of emotion or vague, metaphysical meandering about personhood and morality, or a priori assumptions about moral integrity being, of necessity, only a human capacity. Frankly, I don't mind whether the agent is a human or a robot, as long as morality is respected. Dying at the hands of a robot in war need be no more painful or unjust as dying at the hands of a human. If I am no less a hero for sacrificing myself to prevent a natural disaster as a human wrong, then decorate me for valor in action against a machine just as for action against a human foe. And if an android saves the day, why not reward it with a tune up and a fresh coat of paint?

What I object to is the way the programming is usually assumed to be done in the literature on autonomous systems. Conventional algorithmic models of AI are still the norm in all but a few areas of robotics and autonomous systems. That means that morality and law are coded as rules, actions being initiated or prevented when situations are recognized as instances falling under a rule. But deontology as a framework for ethical decision making has well-known limits. Rules are rigid and inflexible. Situations vary so as to make computationally impossible, for all practical purposes, the identification of every possible individual situation as being of this or that kind, falling under this or that rule. This is why Plato, in *The Republic*, declined to write laws for every occasion, preferring to trust the judgment of good people.

If there is a lesson to be learned from the foregoing discussion for a future in which humans are no longer always in the loop, perhaps it is precisely the lesson that Plato taught over 2,000 years ago and that his student, Aristotle, elaborated into a theory of the moral and intellectual virtues. What might be suggested is a different model for moral artificial intelligence. Perhaps the model should be one in which the agent is equipped with a basic repertoire of moral and intellectual capacities, the ability to learn by adapting those basic capacities to new situations by responding to the morally and cognitively relevant similarities (at first in virtual spaces and under the watchful eye of human or machine monitors), and an ability to learn by emulating the habits of more highly developed moral machines. Much of the technology for doing this is already there in what is known as genetic or evolutionary algorithms,[12] which have proven themselves in computationally intractable areas like the modeling of market behavior and pattern recognition. What is lacking is its deployment in the modeling of moral decision making. If a genetic algorithm can learn reliably to find my face in a variety of poses and circumstances among billions of images on Facebook, or that of bin Laden in the flood of drone video surveillance, why can a genetic algorithm not also distinguish situations in which courage requires resolute action from those in which discretion is the better part of valor?

Argument about a proposal such as this can go on at great length. The technical questions are interesting and complex. The philosophical objections will arouse greater passion. Let me end by speaking to just one of the latter. It will be said that virtue is a uniquely human possession because it develops only among social beings. That virtue characterizes humans in community is axiomatic for Aristotle, mature habit requiring interchange with others for it to be nurtured and maintained, and some of the specific virtues—think again of munificence—make sense only as the agent is oriented toward another. The wild child or the long-time hermit is a creature not of habit but of routine.

Community is crucial, both human community and moral community. But does humanity make the agent moral, or does morality make the agent human? Wisdom is foremost among the virtues for Aristotle, and reason is the faculty that distinguishes humans from the brutes. It would seem, then, that morality in the enlarged sense that includes the intellectual virtues is what makes us human, or is at least what makes us fully realized human beings. Why then cannot a community of machines be judged as moral, and, perhaps in such be judged as "human" as well? What is the residuum of the human if morality be taken away?

[12] In brief, a genetic or evolutionary algorithm begins with a simple program modeled after a genome that controls behavior. It "reproduces" with variation/mutation in the genome. Offspring are set a task, the likelihood of their reproducing, in turn, again with variation/mutation, being dependent upon their success in executing the task. Iteration of this procedure of reproduction with variation and selection over several generations produces sometime surprising variety in approaches to the task and often surprising facility at accomplishing the task. A helpful, recent primer is Eiben and Smith 2010.

References

DePaul, Michael, and Linda Zagzebski. 2007. *Intellectual virtue: Perspectives from ethics and epistemology*. New York: Oxford University Press.

Dewey, John. 1929. *Experience and nature*. 2nd ed. Chicago: Open Court.

Eiben, Agoston E., and J. E. Smith. 2010. *Introduction to evolutionary computing*. New York: Springer.

Kant, Immanuel. 1785. *The groundwork of the metaphysics of morals*. In *Cambridge texts in the history of philosophy,* ed. Mary Gregor. New York: Cambridge University Press (1998).

Koh, Harold Hongju. 2012. International law in cyberspace. Address to the USCYBERCOM Interagency legal conference, Ft. Meade, MD. http://www.state.gov/s/l/releases/remarks/197924.htm. Accessed 18 Sept 2012.

MacIntyre, Alasdair. 1981. *After virtue*. Notre Dame: University of Notre Dame Press.

National Research Council. 2009. *Technology, policy, law, and ethics regarding U.S. acquisition and use of cyberattack capabilities*. Washington, DC: The National Academies Press.

Nussbaum, Martha. 1988. Nature, functioning and capability: Aristotle on political distribution. *Oxford Studies in Ancient Philosophy* 6:145–184.

Nussbaum, Martha. 2000. *Women and human development: The capabilities approach*. Cambridge: Cambridge University Press.

Nussbaum, Martha, and Amatrya Sen, eds. 1993. *The quality of life*. Oxford: Clarendon.

Sen, Amatrya. 1985. *Commodities and capabilities*. Amsterdam: North-Holland.

Sen, Amatrya. 2009. *The idea of justice*. London: Allen Lane.

Shafer Landau, Russ. 2011. *The fundamentals of ethics*. 2nd ed. New York: Oxford University Press.

Vallor, Shannon. 2013. Armed robots and military virtue. In *Ethics of information warfare*, eds Luciano Floridi and Mariarosaria Taddeo, xxx–yyy. New York: Springer.

Chapter 11
Armed Robots and Military Virtue

Shannon Vallor

Abstract This article examines how the accelerated development of semi-autonomous or autonomous armed robots may challenge traditional conceptions of military virtue. While early reflections on the ethical implications of military robotics have focused primarily on utilitarian or deontological/rule-based considerations rather than questions of virtue or character, a comprehensive inquiry into the ethical impact of armed military robots must not ignore the role of military virtue in restraining and legitimizing armed conflict. Armed military robots problematize this role in three ways: first, by potentially leveling the distinction between mere military *action* and virtuous military *service*; second, by possibly diminishing the scope of the military *profession* and its cultivation of military virtue; and third, by undercutting the expectation of virtuous *motivation* in warfare that gives rules of engagement at least some restraining force and helps to distinguish the practice of war from mercenary or criminal violence. I suggest that by initiating or accelerating such shifts in the meaning and practical scope of military virtue, the widespread deployment of armed military robots may have ethically deleterious effects on human soldiers and civilians independently of whether optimistic utilitarian predictions of reduced casualties and collateral damage are realized.

11.1 Section I

The accelerating development and deployment of weaponized military robots has, unsurprisingly, magnified a number of existing social and ethical concerns about the increasingly technical nature of modern warfare. Some of these concerns arise within the military context—for example, the use of armed robots adds another layer of complexity to the ongoing debate about how a modern military ought to balance motives such as force protection, mission accomplishment and minimization of

S. Vallor (✉)
Santa Clara University, Santa Clara, CA, USA
e-mail: svallor@scu.edu

L. Floridi, M. Taddeo (eds.), *The Ethics of Information Warfare*,
Law, Governance and Technology Series 14, DOI 10.1007/978-3-319-04135-3_11,

collateral damage. Other issues are rooted in a wider public context—for example, the political appeal of armed robots as a way to accommodate domestic intolerance for military casualties. Weaponized robots also give rise to a host of new ethical questions: who will we hold legally and morally accountable for the use of military force by robots? Does the potential exist for some form of robotic agency? When, if ever, would it be ethical to give weaponized robots autonomy in targeting and firing decisions? Will robotic forces make war more or less palatable as a solution to political conflict?

Together, these concerns mandate extensive and widespread critical inquiry and reflection on the use of armed military robots; fortunately, this discourse is now well underway (Sparrow 2007; Asaro 2008; Singer 2009; Lin 2010; Sullins 2010; Arkin 2010; Sharkey 2010). Yet there is one subset of these concerns that I will argue warrants increased attention, as it engages matters of crucial moral and civic import that are likely to be less visible to public and political interests than those identified above. These concerns center on the concept of *military virtue*.

Many readers acquainted with the realities of modern warfare may find the concept of military virtue to be of doubtful moral relevance; the 'realist' cynicism of such readers will be informed by My Lai, the Cambodian 'Killing Fields,' Srebrenica and a host of other vicious examples of military atrocity that have taken place since the Second World War. Even well-ordered military bodies that explicitly grant the existence of moral and legal constraints in the conduct of war routinely fail to prevent gross violations of those constraints. Nor are the evils of war limited to shocking atrocities that scar our historical memories; even without these we should not lack for examples of moral viciousness in the conduct of war (Arkin 2009). Needless environmental corruption, cultural devastation and economic exploitation are just a few of the common byproducts of modern military engagements. At this point we may recall with approval Santayana's claim that "To call war the soil of courage and virtue is like calling debauchery the soil of love" (1905, p. 83).

Thus before I can satisfy my primary burden in this chapter, namely, to show how the use of weaponized and even lethal military robots may place significant new pressures on the concept of military virtue, I must first explain why, given the considerations noted above, we should take the concept of military virtue seriously. For I claim that the concept of military virtue is of immense moral relevance today, in spite of, and even *because of*, the evidence that humans remain distressingly prone to egregious violations of its standards. Why? In short, I suggest that rather than serving to *disguise* the moral viciousness of war, cultural ideals of military virtue actually supply an essential part of the contextual background against which the moral viciousness of war is made starkly apparent. If we fail to take military virtue seriously, or if we fail to preserve a practical context for its cultivation and expression, we may further endanger our ability to conceive of warfare as a practice for which humans must be held morally *accountable*.

With Santanyana I reject the claim that war, with all of its cruelty and devastation, is the soil or *source* of military virtue. Ideals of disciplined, discriminating, selfless and humane military character should instead be understood as expressions of a society's refusal to grant the practice of war full immunity from moral standards.[1] Military virtues of courage, selfless service, compassion, loyalty, restraint and discipline are various ways in which moral values *external* to the conduct of war are imposed upon its practice. When such values are formally imposed on military life and practice through the cultivation of virtues within a military *profession*, as discussed later in this chapter, they mark the refusal of a nation or culture to wholly externalize and demoralize its own military conduct. Ideals of military virtue block the displacement of military action to an extra-moral realm to which the norms of that society do not apply. Of course, military virtues are expressed in actions that would often be unacceptable within other social contexts, but this is true of virtues generally—they are highly context-dependent in their expression (Aristotle 1984).

Even so, we must grant that there is a special tension between moral virtues and the distinctive goals and means of military practice, a tension that makes the concept of 'military virtue' intrinsically fragile (there is a useful parallel with the tension that marks contemporary discussions of 'business ethics' in market economies.) As a consequence, some military bodies fail or refuse to cultivate a stable professional identity that incorporates norms of virtue. Even militaries that explicitly cultivate such an identity will struggle to consistently enforce virtuous conduct by their members. This is not, I argue, an indication that 'military virtue' is a confused or impotent concept; rather, it is an indication of the very function of the concept as an expression of cultural resistance to the morally unfettered pursuit of military aims.

This concept underwrites the ability of modern societies to make meaningful distinctions between morally acceptable military aims (national security, the prevention of genocide, etc.) and suspect or flagrantly illegitimate ones (annexation of additional territory, usurpation of wealth or natural resources, extermination of ethnic or religious groups). Furthermore, it allows the evils of war to be seen for what they are, namely, dimensions of human conduct that profoundly wound our moral identities. Though the conduct of war can to varying degrees be restrained by ideals of virtue, war is fundamentally an expression of human immorality. It must not be put aside as a separate domain in which we pretend that we are playing a special, though admittedly nasty, kind of 'game' to which the encompassing demands of moral life do not apply. Far from highlighting the immorality of war, rejecting or ignoring the concept of military virtue makes it just that much easier to instrumentalize, compartmentalize and excuse its evils—to regard them as 'necessary' if also 'regrettable' and 'unfortunate' effects of war, rather than *immoral* dimensions of human conduct for which we are ultimately accountable.[2]

[1] Such ideals are, of course, culture-bound and thus locally and historically contingent in both their strength and expression. While I argue that these ideals serve as a cultural restraint upon military conduct, they are internalized within, and adapted to, military praxis.

[2] In the background here lies an important question: if war is an immoral enterprise, then doesn't its professionalization under the restraint of moral norms lend it an undeserved legitimacy? Rather than incorporate it within our value-system, or allow it to exist outside of that system, would it not be better simply to *reject* it? There is, of course, a long-standing debate concerning the merits and prospects of human pacifism that I cannot engage within the confines of this chapter. Since I regard it as overwhelmingly likely that war will, for the foreseeable future, continue to exist, my argu-

Having provided one *prima facie* reason to take the concept of military virtue seriously, I turn next to the second burden of my argument. I wish to show how armed military robots problematize military virtue in three interrelated dimensions: first, by leveling the distinction between military action and military *service* by virtuous agents; second, by potentially diminishing the scope of the military *profession* and its cultivation of military virtue; and third, by undercutting the expectation of virtuous *motivation* in warfare that gives the rules of engagement at least some restraining force and helps to distinguish the practice of war from armed conflict more generally. I do not assert the inevitability of such effects. I merely wish to draw the reader's attention to the *possibility* that by initiating or accelerating such shifts in the meaning and practical scope of military virtue, the widespread deployment of armed military robots may have ethically deleterious effects on human soldiers and civilians independently of whether optimistic utilitarian predictions of reduced casualties and collateral damage are realized. Whether these deleterious effects would outweigh such utilitarian gains is a matter for another inquiry.

11.2 Section II

The modern concept of military virtue, and its professional cultivation, is most frequently and intimately connected with the concept of *service*. The concept of 'service' in the context of military life has several layers of meaning.

1. It may refer to the length of time that a soldier commits to be at the discretion of a nation's armed forces ('She gave the Navy seven years of service').
2. It may refer to the institution itself ('He was in the service his whole career').
3. It may refer to the act of serving the national interest ('Service to one's country is the highest form of patriotism').
4. It may convey an act or attitude of willing self-sacrifice ('Selfless service' as one of the seven core U.S. Army values) (Miller 2004, p. 202).
5. Finally, it may refer to the professional identity of a soldier ('Thanks to our servicemen and women around the world').

While all of these senses are interrelated and generally imply one another, most of what I say here will focus on the last three senses of 'service'. These three meanings are nearest to the conceptual core of service as a moral ideal—more specifically, a *virtuous ideal*. Moral virtues, we must remember, are not manifested in isolated acts, nor are they simply a matter of one's sincerely professed beliefs, values or convictions. According to the Aristotelian tradition that shapes modern virtue theory in the West, virtues are habituated and rationally informed states of a person's *character* (Hursthouse 1999, Aristotle 1984). They are states of character that predispose one to excellent action in the context of a complete life. Moreover, in both classical

ment begins from that premise, with an aim to understanding how the introduction of armed robots might support or weaken the (admittedly tenuous) moral restraints of professional military virtue.

Aristotelian and Confucian virtue traditions, virtues are states of character highly attuned to the particularities of time, place, social/cultural/political environment and role; virtues not only allow one to live as an excellent human being *qua* human, they also allow one to be an excellent parent, sibling, teacher, ruler, or *soldier*. A moral virtue is therefore both intrinsically personal *and* relational; the virtue resides in the individual, but its meaning is absolutely dependent upon the individual's familial, civic and professional responsibilities to and for others. Thus when I describe service as a virtuous military ideal, I refer not to an isolated act of service performed, but a state of a person's *character* that is habitually expressed in the particular context of military life. In this sense, 'service' is a way of *being* or living that one cultivates—not simply the performance of an external duty.

Armed military robots have profound implications for our conceptual profile of the virtues of servicemen and women, and for the significance of service as a defining feature of ethical military acts and attitudes. To see why, let us start with the third sense of 'service' outlined above, as action that serves the national interest. A robot can serve in this way, if we take the meaning of military service in the narrowest sense, as an act in isolation. However, a robot cannot serve its country in the more robust sense that is linked with virtue. Certainly, a well-designed and implemented robot can promote the interest, and specifically, the security of, a nation. Yet it is not clear how a robot could 'have' a country to serve. A robot can have a physical provenance, of course. Today's military robots are likely to be constructed from parts of diverse national origin, for example, China or India—but we see already the moral irrelevance of such provenance as it concerns the robot's capacity for service. As noted above, moral virtues imply a relational psychology. To 'have' a nation to serve is to be *obligated* to a particular nation. It is to have had one's physical, psychological and moral identity nourished and shaped by that nation's institutions, and to recognize military and other forms of service as ways of honoring those gifts prepared for us by those who formed and maintained those institutions, and ensuring that they will continue to be available for posterity. National service, then, is not a matter of physical provenance nor of being under the direction of some abstract entity—it is a matter of *recognized relationships and obligations* that ought to be honored. Insofar as armed military robots are incapable of recognizing such relationships and obligations, then, they are incapable of national service.[3]

To many readers, this will have been plainly obvious. Why do I stress it, then? Because armed military robots raise the prospect of handing over the most critical *kinds* of military service to entities that are, at least for the foreseeable future, constitutionally incapable of it; this imperils a core ethical value shaping military culture (possibly *the* core value, since loyalty, courage, etc. may be viewed as extensions or implications of ethical service). To make this more apparent, let us turn

[3] Of course, predictions of robust artificial intelligence raise the question of whether robots will, at some point, become capable of genuine moral relations and obligations. We need not assess the merits of those predictions here, as there is widespread consensus that such a scenario lies in the distant future, if even then. I am willing to grant the abstract possibility of this scenario (and hence the long-term contingency of my conclusions). That said, for the foreseeable future, robots will remain capable only of mimicking acts of service in the moral sense employed here.

to the fourth sense of 'service' noted above. That sense emphasizes service as *self-sacrificing*. I assume that, in our concern with current and foreseeable states of military affairs, I need not belabor the point that armed military robots of the sort we can develop today and in the near future do not have the capacity for self-sacrifice—both because they lack a sense of 'self', and because they are incapable of forming sacrificial intent. Of course, a robot can be programmed to place its physical or operational integrity in jeopardy, but in this case it is the programmers who sacrifice something, and what is sacrificed is not themselves but a robot.

Self-sacrifice is a core component of military service because the highest expression of ethical service is that in which one wholly gives oneself for another, or more properly, for many others—one's fellow unit and servicemembers, one's family and friends back home and their progeny, one's fellow and future citizens, and fellow and future members of the human community. The highest sacrifice possible is to give one's life for the security, freedom and well-being of others.[4] While it is evident that armed military robots have no life to sacrifice, one of the notable attractions of their use is to reduce the numbers of human soldiers called upon to do so. In and of itself, this seems to be an unqualified good. To say that total self-sacrifice by a human being is often an act of the highest ethical kind is not to commit oneself to the morally bankrupt conclusion that we must welcome and perpetuate those circumstances in which it is called for. Yet we do need to ask how a vastly successful, robot-enabled reduction of military casualties on a large scale would alter the core meaning of 'selfless service' understood by soldiers today, or the ideal of virtuous service that it informs. The ethical impact of such a shift should be weighed alongside the other moral goods that may be realized by such means.[5]

The need to pose such a question becomes more evident if we consider the potential impact of armed military robots on yet another form of selfless service identified with the military, one that involves a profound and irrevocable sacrifice of a different kind. Military service commits servicemembers to the possibility that that they may be called upon to kill another human being. In doing so, a servicemember sacrifices something of herself as well, no matter how justified or morally necessary the sacrifice. It is not easy to define what is sacrificed in oneself when one chooses

[4] One need not believe that every soldier sacrifices his or her life for moral reasons of this kind; one might think that some do so mindlessly as a result of military 'programming,' while others may never have believed their lives were seriously at risk. To grasp my point, the reader need only grant that *some* soldiers risk or sacrifice their lives for moral reasons of this sort, and that such sacrifices are (or can be) virtuous.

[5] It must be noted that in one sense this moral dilemma is not at all new, only *heightened* by the existence of armed robots. As warfare becomes increasingly technologized and lethal military actions unfold over ever greater distances, the moral weight of such actions becomes increasingly abstract for its practitioners. Yet even technologically advanced militaries still incur significant mortal losses in extended conflicts. Specific decisions about applying lethal force have yet to be deliberately taken out of human hands. I therefore suggest that the potential for widespread deployment of armed robots, especially with autonomous decision power, presents not merely an amplification of the problem of military force at a distance, but a *qualitative* moral problemshift.

to kill a human being, or many human beings, but for an ethical and compassionate person, it undoubtedly is a sacrifice. As noted earlier, not even the constraints of military virtue can wholly eliminate the dissonance between one's identity as a moral soldier and one's identity as a moral person. A parent or spouse whose loved one returns with the ineradicable memory of killing likely understands the gravity of this sacrifice as well. This is another sort of sacrifice that armed robots may soon shoulder for us; and again, in isolation this seems good. Why not spare human soldiers the psychologically devastating consequences of taking a human life? And yet: it is the self-sacrifice made in killing that makes killing in war terrible *for those who kill*, as well as those who die. What will be the impact of developing robots that not only bear arms but deploy them autonomously, as many expect to follow their use under human firing authority (Sparrow 2007; Asaro 2008)? Do we *want* armed robots to take from us one of the choices that force *killers* to bear a significant part of war's moral cost? Will armed robots take away the sacrifice that ensures that when nations go to war, not even the 'winners' get to emerge unscathed?

Certainly, the moral meaning of selfless service is not exhausted by those circumstances that call upon a soldier to willingly risk her own life, or take the life of another. There are many other kinds of sacrifices that servicemembers make: enjoying less time with their spouses, children and other loved ones, delaying their educational or civilian career opportunities, becoming socialized in a manner inconsistent with easy reassimilation to the broader culture. But these forms of sacrifice are somewhat less distinctive; for example, doctors, athletes and parents make some or all of these sacrifices. Perhaps this is all the better; perhaps it is only humane to use robotic technology to free military servicemembers from their historically exceptional burdens. But what are the broader implications for military ethics and human society at large if we do so?

For let us also remember that the concept of military service does not function *only* to acknowledge the moral worth of servicemembers. It also functions as a way to distinguish one class of military actions, one with a special kind of moral character, from others. Not all actions performed in a military context are acts *of* service. Some, and arguably *most*, military actions are performed without any sacrificial intent; these are actions done routinely and mindlessly, or grudgingly, or sadistically, or because they lie on the path of least resistance, will help to impress our peers or superiors, repay their confidence in us, avoid scorn or punishment, get a promotion, settle a score, or any number of other mundane human motivations. Removing humans from the principal arena for selfless service in matters of life and death therefore risks leveling the morally important distinction between the regulative ideal of military *service* by virtuous agents, and mere military *action*.

It is at this stage of our inquiry that the fifth and final meaning of service noted earlier, as an expression of professional identity, becomes most relevant. For we move now from the consideration of particular acts of service to the regulative ideal of service as a standard for a whole life, one that finds expression in a commitment of the person to the military *profession*. How will armed military robots change the meaning of this profession, and the virtues that define its ideal practice?

11.3 Section III

A profession can be understood from two primary points of view. The first is the point of view of the institution that defines the standards of a particular kind of human practice and trains individual members to honor, execute and uphold them. The second is the point of view of the individual who chooses to enter a profession, a commitment which involves not only *accepting* certain institutional standards and practices but also *internalizing* them as an enduring part of one's personal identity. The latter marks the primary difference between a profession and a 'job'. Of course, not all practices that create distinctive standards and identities are professions. Robert G. Kennedy notes that when professionals 'profess' a distinctive identity, they profess *to* a community, promising not only to uphold certain accepted standards and practices, but to do so in service to the public (2000, p. 3). Professionals primarily protect or provide goods of *public* value, such as health, education, justice, or, in the case of military professionals, security (Snider et al. 1999).

Professionalism, then, is an inherently moral notion, one that even in the private economic sphere (doctors, lawyers, etc.) involves a commitment to public *service*. This service is particularly demanding for military professionals, both because of the gravity of the stakes already noted, and because they are expected to weigh the public good far more heavily than their own (Kennedy 2000). From a report on military professionalism by the Strategic Studies Institute of the U.S. Army War College:

> Officers act as agents of society, both individually accountable to them and, as well, serving to strengthen the claim of the service on the affections of the American people…The officer's motivations are noble and intrinsic, a love for his or her craft—the technical and human aspects of providing the nation's security—and the sense of moral obligation to use this craft for the benefit of society…Because of both the moral obligation accepted *and the mortal means employed* to carry out his or her duty, the officer emphasizes the importance of the group over that of the individual" (Snider et al. 1999, pp. 36–37, emphasis added).

Several points are worth noting here. First is the authors' emphasis on officers not merely fulfilling a public duty but doing so with intrinsically moral *motivations*, and beyond even that, having those motivations be *noble*. As the authors note elsewhere, this nobility must be more than a transient impulse but a mark of virtuous *character* (1999, pp. 38–39). Officers need to be "gentle-men and –women", where this term is used in its traditional sense to identify "persons of character, courtesy and cultivation" (1999, p. 38). The military officer is asked to embody an exceptional standard of excellence that may serve as a regulative ideal of virtue for other servicemembers.[6]

Of further note is the authors' emphasis on the link between this moral excellence and the "mortal means" by which officers are called to exemplify it. It is in

[6] This especially critical during those conflicts in which military service is largely performed by conscripted civilians rather than standing armies of professional soldiers.

matters of life and death that military virtue is traditionally demonstrated: "The officer's honor is of paramount importance, derived through history from demonstrated courage in combat" (*ibid.*). The virtuous officer has not just the courage to kill, or even to die for the mission, but also the courage to die simply in order "to preserve the lives of noncombatants" (1999, p. 34). The virtuous soldier/officer is one whose self-sacrificing actions most perfectly honor the public value of the security of human life and liberty, and who by that very exceptional feature not only provides an example of excellence for other servicemembers to imitate, but also secures "the claim of the service on the affections of [a nation's] people" (1999, p. 36).

Before reflecting upon how these ideals of military virtue may be impoverished or narrowed in scope by the deployment of weaponized robots, it is important to confront an objection that is likely to be raised against my analysis, based as it is on the loftier ideals of military professionalism. The objection will go something like this: 'only ivory-tower intellectuals, the historically naïve, or commissioned officers with no combat memory could sincerely view these ideals as realistic standards of military conduct.' The objector will recount a series of egregious violations of such ideals by men and women of every rank in every branch of service, or will point to the military's own studies suggesting that ethical attitudes and behavior in the ranks fall well short of what one might hope, or expect (Conway 2007). Such evidence has led some scholars to suggest that if our goal is simply to improve upon existing ethical performance in war, armed robots may not have too high a bar to clear (Arkin 2009; Lin et al. 2008; Sullins 2010). If all we are concerned about is ensuring that fewer violations of military standards of ethics and international laws of war *transpire*, then whether we can achieve that goal with weaponized robots is an empirical question. Roboticist Ron Arkin (2009, p. 334) suggests that the answer is almost certainly 'Yes'. Noel Sharkey, on the other hand, believes the answer is 'No' (2010, p. 379). Yet I assert that the solution of this empirical calculus, while ethically significant, must *not* be our only concern.

Why not? Why must we also concern ourselves with the moral ideals of military professionalism if they are so rarely realized? For the same reason that we care about *any* profession's ideals. The Confucian philosopher Mencius, responding to the moral realist's cynicism about exacting standards of virtue, reminds us that a "great craftsman does not put aside the plumb-line for the benefit of the clumsy carpenter" (Yearley 1990, p. 48). Even a great craftsman, having human hands and earthly materials, is doomed to fall short of realizing geometrical perfection. Yet it is the practical cultivation of, and motivation by, that ideal which both distinguishes the exemplary artisan from her apprentice and allows the apprentice to be inspired by her example. The meaning of *every* human profession is defined by such ideals, and the military is no different. Even Arkin, who wishes to replace soldiers with lethal robots, acknowledges that "our military aspires to higher ethical performance" than it delivers, and quotes General Douglas Macarthur's assertion that the "sacred trust" of moral service "*is the very essence and reason for [a soldier's] being*" (Arkin 2009, p. 338, emphasis added).

Indeed, without the ideal of selfless service for others, the modern military profession could not exist as we know it, and if it had *no* virtuous ideals, it could not be a profession at all. Arkin's solution to the ethical failures of professional soldiers is to take the responsibility for ethical performance out of human hands, and place it in the hands of robots. Indeed, he suggests that we cannot even trust humans to remain 'in the decision loop' of lethal robotic action, for we are as likely (or perhaps *more* likely) to commit atrocities with these technologies as without them (Arkin 2009, p. 338). This is his argument for the development of artificially intelligent and *ethically autonomous* lethal robots. Yet as Joanna Bryson and Philip Kime have noted, our culture's current obsession with artificial intelligence as a solution to human shortcomings can obscure the fact that it is not simply the *results* of actions that we value, but the performance of actions themselves (Bryson and Kime 2011). Arkin seems to focus exclusively on the negative pole of moral valuation, e.g., the duty to *avoid* the occurrence of human misdeeds. Surely this should be one of our central aims. But it would be profoundly mistaken for us to ignore the positive need humans have for *accomplishment* of moral actions—and it is not clear how much room for that Arkin is willing to leave us, at least in the domain of military life.

This is not to say that the introduction of armed robots must deprive the militaries that rely upon them of *all* virtuous ideals. Yet insofar as robotic agents of lethal force reduce the practical scope of expression of selfless service, which currently functions as the *primary* regulative moral ideal of modern militaries, the ethical impact of armed robots could be of profound significance. Furthermore, if robots can be trusted to be better ethical performers in the domain of lethal action, why not hand over to them all *other* military responsibilities in which humans have a spotty ethical record? *If* robots with electronic 'ethical governors' can be more ethical than humans in killing (Arkin 2009), then surely they can be more ethical in many other contexts, and if they can do it all better, why not let them? In the extreme case, then, the remaining scope for human expression of military virtue could be vanishingly small. Yet in practical terms that scenario is not a present worry. It is the realm of lethal action that is the more proximate concern, as Arkin's own project suggests.

The realm of lethal action is also the most philosophically significant in terms of military virtue. In making his case against the use of lethal armed robots, Noel Sharkey (2010) quotes a military colonel, Lee Fetterman, who states that:

> Men should decide to kill other men, not machines. This is a moral imperative that we ignore at great peril to our humanity. We would be morally bereft if we abrogate our responsibility to make the life-and-death decisions required on a battlefield...This is not something we would do. (2010, p. 380)

What is interesting is that Sharkey uses this quote to support his argument that robots are *incapable* of meeting the legal and ethical requirements of proportional and discriminating use of military force (2010, p. 379). He may well be right. Yet it seems to me that Colonel Fetterman's words address quite a different matter, and that his claims would not be affected by the success or failure of Sharkey's argument. Colonel Fetterman's claims are about what good soldiers cannot, or *would not*, do. Let us momentarily assume, with Arkin and against Sharkey, that robots *can* be developed that meet the relevant ethical and legal standards. This would be a fact about what *robots* can do. What does this have to do with *our* moral identity and sense of duty as human beings, or as professional soldiers—why should a fact about what robots can do change this officer's sense of what a good human soldier would *not* do? Thus I suggest that we take pains not to conflate, as is too easily done, two important philosophical concerns:

1. Whether autonomous and ethical armed robots are empirically possible;
2. What the use of armed autonomous robots would mean for the moral character of human soldiers and the modern military profession.

What should we conclude about armed robots and modern military professionalism? The accelerating pace of development of armed robots engages ongoing military debates about the relative priorities of 'force protection', mission success and minimizing civilian harm (Pfaff 2011). It remains to be seen whether wide deployment of armed autonomous robots can provide vastly improved civilian *and* force protection without compromising mission success. But let us assume optimistically that it will. We are then faced with a number of important questions. If robotic means of force protection were to vastly reduce the scope and moral gravity of the sacrifices expected of members of the military profession, how would this change how the profession understands and regulates itself? How would the moral obligations of soldiers to serve the public selflessly be reinterpreted? How would this impact military efforts to recruit, motivate and socialize its members to cultivate a distinctive professional and moral identity? With fewer opportunities for exemplary acts of sacrificial service, would the military continue to secure the affections of the public that it serves? Would we begin to think of virtue less as something to be cultivated, and more as a software parameter to be calibrated? Would military excellence be reduced to little more than a special form of engineering excellence? Are the professional virtues presently valued in engineering schools and R&D labs at all sufficient to give those who design, program and supervise armed robots the ability to understand and respond wisely to the moral gravity and complexity of war?

11.4 Section IV

One may wonder whether the impoverishment of moral meaning within the military profession may be a good thing. Perhaps weaponized robots will strip the veneer of high moral service from the ugly realities of war, allowing us to recognize war as something not deserving of our sacrifice, or our virtues. Perhaps we will finally find better ways of resolving our internecine conflicts. This, however, is not the most plausible outcome. For humans have never restricted themselves to forms of violence that purport to be moral. Human history is pervaded by other forms of violence, of which two should concern us here—violent *crime*, that is, violence that breaks legal boundaries established by society, and *mercenary* violence—violence that takes place within ostensibly legal boundaries but is motivated by private profit rather than public service. The impoverishment of ideals of military virtue is less likely to lead to peace than to the further conflation of professional military action with morally unfettered violence, and the continued encroachment of criminal and mercenary violence in modern armed conflict.[7]

This is highly pertinent to the increasingly *asymmetrical* nature of modern warfare (Arquilla 2011; Thornton 2007). The recent rise of underfunded non-state actors engaging in international conflict with large, technologically advanced national militaries generates a host of strategic, cultural, political and moral quandaries. But let us add just one more. How will the military deployment of weaponized robots affect the behavior, perceptions and beliefs of combatants without access to these technologies?[8] To do justice to this question, we must shift the emphasis of our inquiry away from the role of military professionalism in shaping the motivations, conduct and identity of a nation's servicemembers, to its role in shaping the attitudes, beliefs and conduct of those beyond its authority—that is, foreign combatants, non-combatants and the world community.

We noted above that it is the moral meaning of military service, and the professional identity it fosters, which allows a people to distinguish its servicemen and women who deserve affection, honor and gratitude from mere mercenaries or criminals, who are more commonly despised. It also plays *some* significant role in the capacity of foreign publics to distinguish between *soldiers* of an opposing force and invading 'barbarians' (Singer 2010). Of course, other factors are heavily at work in such dynamics: the nationalist rhetoric of political, cultural and religious authorities, a people's historical memory of war and its humiliations, and the very primitive, entirely reasonable fear of the Other who arrives at your doorstep bearing arms. Still, there exists a rich spectrum of possible responses to a foreign enemy, ranging from passive surrender, to vigorous opposition by recognized military means, to organized resistance by any conceivable means, and finally to disorganized, leaderless panic and violence, in which distinctions about means *and* targets are entirely given up to moral chaos.

Historically it is the latter half of the spectrum that often spawns the most egregious offenses against humanity, on *both* sides of the conflict. It is also that end of the spectrum that renders military strategy least effectual, verging on pointless. It would seem, then, that if armed conflict remains a distinctive feature of human existence, it is in the interest of armed forces to take pains to steer the defensive response of hostile peoples to the part of the spectrum that (if passive surrender be a vain hope) produces an organized counter-response of a professional military.

[7] The notorious involvement of private contractor Blackwater USA (now xE Services) in Iraq is instructive here. See Singer (2003) on the broader trend of state-sponsored mercenaries.

[8] It is true that such combatants are already exploiting opportunities to develop small-scale military robotics of their own. Still, an *ad hoc* remote-control truck or model airplane with a grenade launcher bolted onto it is hardly symmetrical with the powers of weaponized robots that could be generated by a military research laboratory with a billion-dollar budget. It is hard to see how underfunded combatants without the support of a strong state military could afford to develop or deploy accurate and reliable weaponized robots on a large scale.

Will armed robots make that more or less likely? The answer is not a simple one; it depends upon a large number of contingent factors pertaining to particular conflicts, factors that cannot be entirely calculated or even known in advance. Still, some general observations might be worth making. One interesting feature of professions is that because they typically promote goods that are recognized as having significant and enduring human value, they tend to permeate cultural and political boundaries; for example, we can expect to find physicians, educators, and soldiers in every minimally stable society. Typically, one also finds a level of mutual recognition between professionals of the same kind that crosses those same boundaries. Of course, this recognition may not always be accompanied by esteem, and may break down entirely when standards of professional practice diverge too far; consider the absence of reciprocal professional recognition between tribal doctors of sixteenth century North America and their European counterparts.

Still, most professions *do* share certain core standards of virtue across political and cultural boundaries. An exemplary French lawyer expects an exemplary Japanese lawyer to have a strong respect for the rule of law, an excellent memory for detail, and argumentative skill. An exemplary Indian military officer expects an exemplary Australian officer to reliably display courage, discernment, decisiveness, calm under pressure, chivalry, leadership, patience, mental, emotional and physical discipline. Unlike standards of virtue that are narrowly tied to a certain cultural or national identity, core professional standards are also rooted in the distinctive identity of all who commit to secure for the public the particular human goods associated with that profession. I suggest that this mutual recognition of morally motivated service *promotes* (though it cannot assure) a certain restraint *by* those standards among exemplary members of a profession, even when their aims conflict. I will argue that in some contexts, and particularly in asymmetrical conflicts, foreign combatants may not recognize forces using armed robots as morally motivated members of the same professional practice—potentially making it less likely that those combatants will be restrained by international standards of ethical combat and the laws of war associated with them.

As P.W. Singer has noted (2010, p. 309), where modern militaries have cultivated a professional identity around a moral ideal of virtuous service, one function of this ideal has been to distinguish between *legitimate* military actions and actions which, even if done under the authority of military agents, fall into the realm of criminal or mercenary violence. If the modern ideal of virtuous military service *at present* entails personal exposure to the mortal costs of war, then a foreign combatant who recognizes himself as a member of a professional military force may be unable to recognize combatants who apparently evade those costs by fighting with armed robots *as* morally motivated military professionals. Here I am raising a possibility that is largely speculative, though one for which there is at least some anecdotal support (Singer 2009, 2010; Lin 2010). I am not asserting the objective validity of such a perception, only the plausibility of its existence. Persons or groups with this perception are arguably more likely to view themselves as facing an invasion of criminals, mercenaries or 'techno-barbarians' rather than an incursion of professional *soldiers* serving their nation.[9] Of course, even a professional military must expect vigorous resistance from those subject to its force. Yet when soldiers are recognized *as* soldiers *by* soldiers, and by civilians who understand the shared moral motivations and virtues of the military profession, that opposition is, I claim, more likely to be restrained in its *means* of resistance by internationally recognized standards of lawful and ethical combat. *If* this is true, and up to this point I have offered no more than a speculative basis to think it might be, then militaries that employ armed robots may in fact undermine global peace and security, including those of their own peoples, by unwittingly encouraging their enemies to abandon ethical restraint with respect to the acceptable means and targets of armed conflict.

Even if nations that rely upon armed robots find new ways for their soldiers to express the distinctive military virtues of selfless service, courage and so on, these will likely be less visible to foreign combatants in nations that still rely primarily on more traditional forms of military self-sacrifice. It is thus in *asymmetrical* contexts where the problem of mutual professional recognition and restraint is most likely to arise—yet ironically, technologically advanced militaries are drawn to the development and deployment of armed robots primarily *as a response to* the challenges of increasingly asymmetrical and irregular conflict (Singer 2009, p. 221).

Of course, when dealing with insurgents who bomb public markets and litter roads with IED's, it may seem absurd to talk about restraining the use of technology in order to uphold conventional military norms of courage and self-sacrifice. And perhaps it *would* be, if those insurgents were the only persons to whom our military practices have meaning. But it is precisely the attitudes and perceptions of civilians, local officials and regular military and security forces that are often the key to weakening an insurgency. In Afghanistan and Iraq, it was not committed insurgents whose loyalties U.S. and allied troops were primarily trying to influence; it was tribal leaders and others who may or may not continue to back the insurgents. If one's forces come to be widely perceived as 'techno-barbarians' lacking the military virtues of courage and selfless service, the task becomes that much harder.

The problem is not necessarily permanent. Militaries that employ the protection of robotic arms may not be universally perceived as professional or virtuous today, but this could well change. Consider, for example, the introduction of technologies such as rifles, high-altitude bombers, and crossbows (Lin 2010) which, by making the application of lethal force more remote and less risky, challenged the same military ideals of virtuous courage and self-sacrifice as armed robots. It took time to develop professional roles around these technologies that could be perceived as

[9] The issue here is not simply one of risk exposure; it is about a recognizably professional context of practice. Why is a military sniper respected, while a poisoner is not, even if they have the same target? Both typically operate with *some*, though not total, remove from personal risk. But the sniper is a professional; he is not merely capable of completing a specific act, he occupies a well-defined role with normative standards and virtues recognized by his fellows; the poisoner is typically an isolated outlier, not apprenticed to a role within a practice. However, this is a contingent matter – we could imagine militaries coming to cultivate and mutually recognize professional ranks of poisoners, while snipers, at the invention of the rifle, were often seen as less virtuous fighters, especially by enemy ranks without comparable professional roles.

consistent with those norms. There may be a qualitative difference, however, in the case of autonomous armed robots that remove, rather than merely distance, human soldiers from the physical and psychological costs of applying lethal force.

Whether or not this is so depends in part upon whether ideals of military virtue can be adapted to this practice, and whether sophisticated military robotics become more than the exotic tools of a few wealthy nations. Greater symmetry of use and a cultivation of shared standards of virtuous robotic practice could lead humans who fight with the aid of robots to be widely viewed as morally motivated 'professionals'. There are other optimistic scenarios: the deployment of robots not as lethal agents but as military supervisors or 'ethical police' has been suggested as a way of ensuring greater human professionalism on the battlefield (Arkin 2009). Or perhaps the cultivation of military virtue in human soldiers will be advanced by the force amplification enabled by armed robots, allowing militaries to be more selective in their recruitment and promotion of human soldiers.

Yet none of these optimistic scenarios are foregone conclusions. Consider perhaps the first test case for my thesis: the drone pilot. Even within the military, this role is only beginning to be cultivated as part of an integrated professional practice, and it is not now one that every enemy, or even the very publics those pilots serve, can reliably recognize as such (consider a youth T-shirt sold by American online retailers with the slogan 'Real Men Don't Pilot Drones'). Should a professional practice with a recognizable moral meaning fail to consolidate around the use of drones, or armed robots more generally, then we may see a very different and catastrophic result of their widespread use: the collapse of an already fragile international recognition of the conduct of war as a *professional practice* bound by ethical norms. The moral nightmare of a technologically advanced but unregulated 'free-for-all' in global conflict is the worst-case scenario for humanity, one that is by no means unprecedented or unthinkable in the modern age (Singer 2009; Alach 2011). It may be more militarily prudent, then, for technologically affluent nations to engage the international community in a discussion about ethical and legal standards of weaponized robotic warfare *before* deploying such means on a wider scale, *especially* in asymmetrical/irregular conflicts.

Some might find my analysis here superfluous, since with respect to autonomous lethal weapons, they believe that principles of just war, especially considerations of *jus in bello*, already preclude their use (Sparrow 2007; Asaro 2008). This may be so. But it must be noted that such principles, and the international laws of war founded upon them, already presuppose what I have characterized as a mutual but fragile understanding of war as a shared human *practice*, one governed by the particular moral virtues and standards associated with the military profession. If the deployment of armed robots undermines that understanding, something I have claimed is a distinct possibility, then just war considerations (however valid in principle) will be practically and motivationally compromised as guides to military restraint. As George Lucas suggests, it is not unreasonable to think that this may lead to what he calls the "de-valorizing" of war, along with what he and others worry will be the lowering of human resistance to engaging in it (Lucas 2010, p. 296).

11.5 Section V

I have argued that an inquiry into the ethical implications of armed military robots must include careful reflection upon the moral meaning of the military profession, its ideals of selfless public service and the standards of virtuous practice through which those ideals get expressed and recognized in professional military action. I have suggested that the development and deployment of armed military robots on a wide scale may problematize the modern military ideal of selfless service, and may displace the traditional virtues of military character that have historically been cultivated within the armed forces to promote the effective realization of that ideal. I have claimed that this displacement may also lead to a decline in, or a reduction in the scope of, modern military professionalism. Finally, I have suggested that forces that deploy weaponized robots may exacerbate asymmetrical or irregular conflicts by undermining the recognition of their own forces as members of a common military profession bound by shared ethical and legal restraints.

Nothing in my analysis entails that there is no role for robots on the battlefield; with respect to the performance of many of those military tasks commonly classified as 'dull, dirty and dangerous' (Lin 2010), there are obvious instrumental and even ethical benefits to the development and use of military robotics. Nor does my argument demonstrate that the use of autonomous lethal robots is logically incompatible with the preservation and continued cultivation of professional military virtue—I take my argument to show only that: (1) there are good reasons to worry that the use of the former will have a deleterious impact on the latter, and (2) that the latter is of sufficient ethical importance that ethicists and military professionals ought to take this worry quite seriously.

One final note: in addition to encouraging ethical discipline in the conduct of war, I have argued that ideals of military virtue also play a significant role in how soldiers are perceived by the publics they serve, and by foreign combatants and civilians. They also, quite obviously, influence how soldiers perceive themselves. Several scholars have emphasized the need to understand how armed robots will impact the psychological and moral experience of human soldiers and teleoperators (Sparrow 2009; Singer 2009; Sullins 2010; Sharkey 2010). The virtuous ideals of selfless service perform a psychological function as well as an ethical one. While they can never insulate a soldier fully from the psychic costs of engagement in a business as violent as war, they help to moderate those costs by embedding them in a moral context of virtuous self-sacrifice for the public good. We must ask whether the deployment of armed robots, by eroding that context, actually risks the disintegration of the soldier's ethical self and the exacerbation of social and psychological dysfunction in the military ranks. While such a worry is highly speculative, and depends upon the particularities of that deployment, it must not be dismissed out of hand. It would be the worst sort of irony if a technology intended to detach soldiers from the most psychologically costly duty of their service in fact detached them from the very moral identity that makes that service bearable.

References

Alach, Zhivan. 2011. The new Aztecs: Ritual and restraint in contemporary western military operations. U.S. Army War College/Strategic Studies Institute, http://www.strategicstudiesinstitute.army.mil/pubs/display.cfm?pubID=1076. Accessed 27 Nov 2011.

Aristotle. 1984. *The complete works of Aristotle: Revised Oxford translation*, ed. J. Barnes. Princeton: Princeton University Press.

Arkin, Ronald C. 2009. *Governing lethal behavior in autonomous robots*. Boca Raton: CRC Press.

Arkin, Ronald C. 2010. The case for ethical autonomy in unmanned systems. *Journal of Military Ethics* 9:332–341.

Arquilla, John. 2011. *Insurgents, raiders, and bandits: How masters of irregular warfare have shaped our world.* Lanham: Ivan R. Dee.

Asaro, Peter. 2008. How just could a robot war be? In *Current issues in computing and philosophy*, eds. P. Brey, A. Briggle and K. Waelbers, 50–64. Amsterdam: IOS Press.

Bryson, Joanna J. and Philip P. Kime. 2011. Just an artifact: why machines are perceived as moral agents. In: *Proceedings of the 22nd international joint conference on artificial intelligence, 1641–1646*. Barcelona: Morgan Kaufmann.

Conway, James T. 2007. *Mental health advisory team (MHAT) IV brief.* U.S. Army Medical Department. www.armymedicine.army.mil/news/releases/20070504mhat.cfm. Accessed 22 Nov 2011.

Hursthouse, Rosalind. 1999. *On virtue ethics*. Oxford: Oxford University Press.

Kennedy, Robert G. 2000. Why military officers must have training in ethics. *International Society for Military Ethics*. http://isme.tamu.edu/JSCOPE00/Kennedy00.html. Accessed 22 Nov 2011.

Lin, Patrick. 2010. Ethical blowback from emerging technologies. *Journal of Military Ethics* 9:313–331.

Lin, Patrick, Abney Keith and Bekey George. 2008. Autonomous military robotics: Risk, ethics and design. U.S. Department of Defense/Office of Naval Research, http://ethics.calpoly.edu/ONR_report.pdf. Accessed 22 November 2011.

Lucas, Jr. and R. George. 2010. Postmodern war. *Journal of Military Ethics* 9:289–298.

Miller, J. Joseph. 2004. Squaring the circle: Teaching philosophical ethics in the military. *Journal of Military Ethics* 3:199–215.

Pfaff, Tony. 2011. Resolving ethical challenges in an era of persistent conflict. U.S. Army War College/Strategic Studies Institute. http://www.strategicstudiesinstitute.army.mil/pubs/display.cfm?pubID=1058. Accessed 22 Nov 2011.

Santayana, George. 1905. *The life of reason: Reason in society*. New York: Charles Scribner's Sons.

Sharkey, Noel. 2010. Saying 'no!' to lethal autonomous targeting. *Journal of Military Ethics* 9:369–383.

Singer, Peter W. 2003. *Corporate warriors: The rise of the privatized military industry*. Ithaca: Cornell University Press.

Singer, Peter W. 2009. *Wired for war: The robotics revolution and 21st century conflict*. New York: Penguin Group.

Singer, Peter W. 2010. The ethics of killer applications. *Journal of Military Ethics* 9:299–312.

Sparrow, Robert. 2007. Killer robots. *Journal of Applied Philosophy* 24:62–77.

Sparrow, Robert. 2009. Building a better WarBot: Ethical issues in the design of unmanned systems for military applications. *Science and Engineering Ethics* 15: 169–187.

Sullins, John. 2010. RoboWarfare: Can robots be more ethical than humans on the battlefield? *Ethics and Information Technology* 12:263–275.

Snider, Don M., John A. Nagl and Tony Pfaff. 1999. Army professionalism, the military ethic and officership in the 21st century. U.S. Army War College/Strategic Studies Institute. http://www.strategicstudiesinstitute.army.mil/pubs/display.cfm?pubid=282. Accessed 22 Nov 2011.

Thornton, Rod. 2007. *Asymmetric warfare: Threat and response in the 21st century*. Cambridge: Polity Press.

Yearley, Lee H. 1990. *Mencius and Aquinas: Theories of virtue and conceptions of courage.* Albany: SUNY Press.

Chapter 12
Deception and Virtue in Robotic and Cyber Warfare

John P. Sullins

Abstract Informational warfare is fundamentally about automating the human capacity for deceit and lies. This poses a significant problem in the ethics of informational warfare. If we want to maintain our commitments to just and legal warfare, then how can we build systems based on what would normally be considered unethical behavior in a way that our commitments to social justice are enhanced and not degraded by this endeavor, is there such a thing as a virtuous lie in the context of warfare? Given that no war is ever fully just or ethical. And that navigating the near instantaneous life and death decisions necessitated by modern conflicts fully taxes the moral intuitions of even the best trained and well intentioned war fighters. It follows, that we need accurate analysis on whether or not we can construct informational technologies that can help us make more ethical decisions on the battlefield. In this chapter I will focus on the fact that robots and other artificial agents will need to understand and utilize deception in order to be useful on the virtual and actual battlefield. At the same time, these agents must maintain the virtues required of an informational agent such as the ability to retain the trust of all those who interact with it. To further this analysis it is important to realize that the moral virtues required of an artificial agent are very different from those that are required of a human moral agent. Some of the major differences are that a virtuous artificial agent need only reveal its intentions to legitimate users, and in many situations it is actually morally obliged to keep some data confidential from certain users. In many circumstances cyber warfare systems must resist the attempts of other agents, human or otherwise, to change its programming or stored data. Given the specific virtues we must program into our cyber warfare systems, we will find that while human agents have many other drives and motivations that can complicate issues of trust, we will find that in comparison to human agents, artificial agents are far less complex and morally ambiguous. Thus it is conceivable that artificial agent should be actually more successful at navigating the moral paradox of the virtuous lie often necessitated by military conflict.

J. P. Sullins (✉)
Sonoma State University, Rohnert Park, CA, USA
e-mail: john.sullins@sonoma.edu

L. Floridi, M. Taddeo (eds.), *The Ethics of Information Warfare*,
Law, Governance and Technology Series 14, DOI 10.1007/978-3-319-04135-3_12,

12.1 Deceit in Warfare: Dastardly Behavior or Tactical Brilliance?

The ethics of lies in the context of warfare might seem deeply dependent on context. If we assume a strategic or "realist" framework for our ethical decision making, then from the standpoint of one engaged in a deadly struggle, lies are wrong when you or your allies are the victim but may be correct or even obligatory if you or your allies perpetrate the falsehood and a more just political situation obtains because of it. For instance, under this kind of thinking it was wrong for the Japanese to cloak their attack on Pearl Harbor but right for the US to hide the development and deployment of the atomic weapon, assuming that it was wrong for the Japanese government to have started the conflict with the United States but correct for the US to do everything in its power to end the conflict.

With the notable exception of relativism, most ethical systems are much more circumspect when it comes to the propagation of falsehood. For instance, a strict deontologist would argue against deceit, even when it advanced one's immediately perceived interests even if those interests appear virtuous to the actor. Other systems would allow for very limited forms of deceit, more or less, depending on the situation and or the motives of the active agent.[1] For instance a rule utilitarian might be able to support a rule that allows for one to lie when dealing with hostile agents, especially if that lie might eliminate, impede or damage those agents and result in a situation that maximized the values of the particular utilitarian approach espoused by the moral agent in question be that happiness, human flourishing, or adherence to some set or rule utility.

Here we see the flaccidity of trying to approach this problem with the tools of early modern ethical systems. There is no widespread agreement on whether or not it is permissible for ethical agents to be strategically deceitful when they find themselves in dangerous situations. It just depends on what ethical system you chose, some will allow for it while others will not. Professional philosophers become more or less comfortable with these kinds of systematic impasse and dig their heels in deep and defend their particular flavor of one of these systems to the death. But those outside of philosophy are often deeply troubled by the irreconcilability of the major ethical theories and use this paradox as an indictment the entire project of moral philosophy. The philosopher Eric Dietrich has noticed this fact and has argued that it might be beyond human cognitive capabilities to ever move beyond this deadlock and that it is indicative of deeper flaws in the human ability to undertake the task of philosophy in general (Dietrich 2011a). Interestingly enough, Dietrich is not as pessimistic about the possibility of artificial agents that could move beyond the vexing cognitive limitations of human moral agents and he argues in his essay, "Homo Sapiens 2.0 Why We Should Build the Better Robots of Our Nature," that as humans the one and only truly moral action we can achieve would be to help bring

[1] If one holds the view that there is no truth period, then that certainly ends the discussion. For the sake of having something to say I will not address this possibility in this paper. But as we will see, the strict referential truth-value of a statement may be divorced from its effects on moral agents.

these agents to life and then get out of their way so they can proceed to untangle the moral Gordian knot we have tied around ourselves (Dietrich 2011b). Even if it is possible that future artificial agents might make better moral agents, we are still left with the problem of how to design and program artificial moral agents.

There are a growing number of philosophers engaged in theorizing about the possibility of artificial moral agency and or Machine ethics (see, Anderson and Anderson (eds.) 2011; Lin et al. (eds.) 2011; Sullins (ed.) 2011a; Wallach and Allen (eds.) 2010). But these ideas have yet to be fully expressed in actual technologies. One notable exception to this is the work of the roboticist Ronald Arkin of Georgia Tech. Arkin has been researching technical means of providing some ability for artificial weapons systems to reason on their own about whether or not their actions on the battlefield are remaining in accordance to international standards and laws of conduct in war. As part of that work he has developed the initial designs for an "ethical governor" which is a program that monitors the actions of the weapons system as it autonomously patrols the battlefield and seeks to keep the system from straying outside of programed constraints for the system, much like a governor in a mechanical system keeps that system within safe operating parameters (Arkin 2009a, 2009b, 2010; Arkin et al. 2012). As an example, if the system was engaged with some enemy combatants, the weapons targeting systems would be finding and engaging targets, but this ethical governor would monitor these actions and if the situation changed such that there became too much of a possibility for unacceptable damage to civilians or property, or that the system might need to be constrained due to certain rules of engagement or laws of war that were in effect for this mission, then the ethical governing system would take control of the machine and cease firing (ibid.). This is just one of many conceivable systems but what is most interesting here is that the work is not just theoretical. Arkin and his colleagues are approaching this problem as engineers who are working to develop real systems and products, they see ethics as a kind of technology or at least as something that can be expressed through technology. This move was presaged early last century by the philosopher John Dewey who argued that traditional ethics and morality were incapable of adequately confronting the vexing moral issued raised by the new challenges of a global technological society and he argued that they should be reconstructed as a means for determining new methods for improving value judgments (see, Dewey in Gouinlock (ed.) 1994), and the Dewey scholar Larry Hickman argues that this process can be seen as an instrumental or technological approach to ethics (Hickman 1990). Values are a kind of tool that helps guide conduct and these can be revaluated on the basis of empirical evidence gained while operating under the values in question thus allowing for a kind of moral progress as old values confront insurmountable challenges and are replaced by new ones as was required by the great social changes and conflicts that constituted the era Dewey lived in. In this way one is not appealing to a fixed set of norms or some metaphysical *telos* to make moral judgments but rather a society holds to a developmental set of norms that are always open to revision if they confront a serious challenge that they are unable to otherwise successfully mitigate. This instrumental approach to ethics was further clarified by Mario Bunge who recognized that moral statements were often in the form of conditionals and that could be transformed into a more precise logic or programed into a kind of information technology he called "Technoethics" (Bunge 1977). We can see this approach to ethics mirrored in Arkin's work though it seems that this is a coincidence and not by design. In this paper we will focus on this method of approaching ethics and morality as it allows for us to move beyond the meta-ethical road blocks posed by traditional moral systems that may otherwise prevent work on the specific moral issues that confront machine ethics. We can now return to the more focused discussion of the proper role of deception in artificial systems.

12.2 Information Content and Ethics

Being deceitful is wrong. This is the initial moral intuition that comes quickly to mind. No society can tolerate lies about important matters and no individual wants to live in a world where one cannot trust others to be truthful with them. A lie is a known falsehood masquerading as truth and knowing the truth is always better than being deceived by something that is false. In this line of reasoning we have been speaking of deceit in terms of only true and false, right and wrong. Another option would be to deny the claim that any particular statement is exactly either the truth or a lie. In fact after a bit more contemplation we can see that it is possible to be deceitful while only uttering true sentences. For instance, a sentence may be strictly true but through omission can still mislead other moral agents. As an example one might ask a local informant if there are any enemy combatants in a particular area. To which the informant truthfully answers "no" but also knowing full well that they intend to return soon but lets the interrogator continue on into the area as if it were safe. On the other hand, there are instances where statements that on appeal to factual referent are patently false, yet may still lead to morally beneficial situations. I am thinking here of the "Platonic" lie, or statements like, "my love for you is endless." The former is a paternalistic falsehood that is delivered in an attempt to help the one deceived not suffer unnecessarily or to question something they are incapable of understanding, while the later example is a promise that is improbable in the extreme but a lovely sentiment nonetheless and emotionally very satisfying to the person it is spoken to.

Thus, technically speaking, information content can have a true referent and beneficial consequences, a true referent and maleficent consequences, a false referent and beneficial consequences, and finally a false referent and maleficent consequences (see Fig. 12.1). Therefore, in any give situation an agent must determine the correct amount of truth and falsehood needed to produce a beneficial consequence. It follows then that a moral agent, who wishes to produce beneficial situations, may be called upon to knowingly deceive.

Even this more complex notion of deceit is not entirely adequate. We have so far dodged the question of what makes a particular consequence beneficial or maleficent. Again we seem to be lost in a conceptual muddle as something that seems beneficial to me, such as my gaining access to your savings account, might from

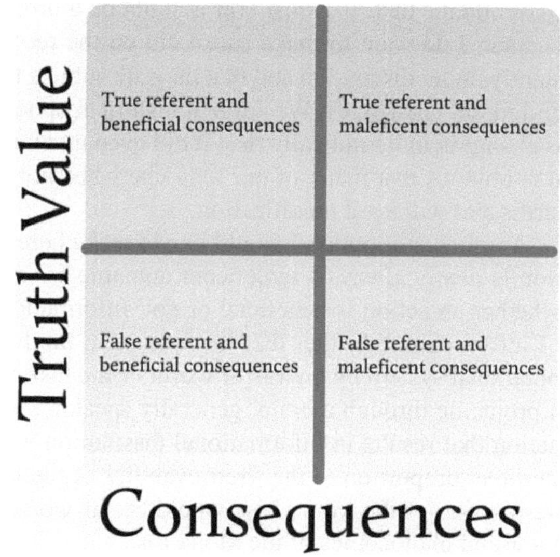

Fig. 12.1 Information content and ethics

your point of view be quite bad. There are, of course, longstanding and venerable arguments in philosophy that attempt to tie this judgment of value to some kind of universal form or numina which can serve as the final arbiter of the beneficial value of some given action. Unfortunately, these titanic debates have yet to resolve into satisfactory answers that are useful for the design of artificial moral agents.

Luckily for us, in the particular context of this discussion we will be able to work around this problem. For now, it is not necessary to attempt to resolve these heady multi-generational debates here as we are engaged in a very well defined arena of discussion. Robotic or cyber warfare (Informational warfare for short), as a subset of warfare in general has a set of very specific set of rules, laws and codes of conduct associated with it which has been developed through international negotiations. On the theoretical side as well a very through philosophical analysis in the form of just war theory has evolved over millennia that serves to inspire moral conduct in the declaration of and execution of war.

Briefly put, just war theory imposes duties on those who would start or fight in wars. Wars can only be propagated by dully constituted authorities whom are motivated by right purpose and only as a last resort to all other means of avoiding conflict (Jus ad Bellum). While warfighters must hold all noncombatants immune to violence while making sure to be proportionate in the violence they can justly impose on enemy combatants and these actions must produce more good than harm (Jus in Bello).

While it would be impossible to argue that these rules, laws and norms are perfect, they are reasonable and provide a good place to ground our discussion. So, for the purposes of this paper then a beneficial situation will be defined as one that does not strongly contradict just war theory or the international rules and law of war.

While the theory of Just War will not be thoroughly questioned in this particular chapter, I do want to make sure I am on the record in advising that they are constantly under discussion and that they are subject to change in light of new evidence and moral challenges. We are in a historical epoch in which the technology of war is changing more radically than it did even under the introduction of gunpowder, so it is obvious that many of our long cherished notions of just warfare will be under stress and will need modification.

Another assumption I would like to make here is that that informational dissolution is nearly always a maleficent outcome for any action and a good measure of whether an action is beneficial or not. Informational dissolution is simply the loss of information, whether that loss comes in the form of the annihilation of a computational system by a virus or worm or the loss of life and memory occasioned by a projectile through a brain, generally speaking both of these are a bad thing. Any action that results in informational dissolution will receive its negative evaluation in direct proportion to the irretrievability of the information lost. For instance, the destruction of the Mona Lisa would be far worse than the destruction of one of a thousand photocopies of the Mona Lisa.

When discussing information in this way it is important to be clear about what is meant by the term "information." Paradoxically, there is no completely satisfying answer to what information is, though the term is obviously very useful nonetheless. Here information is meant is a way that is a bit more philosophically stronger than the way one might define information in an engineering context. Engineers will be happy to define information in the manner of Claude Shannon who describes it as a "signal" which is the ordered set of symbols that can be communicated between two or more agents along some channel with little or no "noise" or loss in the accuracy of the original message (Shannon and Weaver 1949). In addition to this definition of "information" we need to add here a more deeply ontological claim. Information is also something that either constitutes or is very closely correlated with existence itself. This is the basic intuition that motivates the emerging fields of information philosophy and information ethics (see, Floridi 2011). We are straying close to another metaphysical wormhole here as if we take these propositions seriously, then given that everything is constituted of information, it would seem that all warfare is informational warfare. As interesting as that idea is, let's just back away from it for now and return to the more prosaic understanding of information. This way we can see that without overriding moral arguments, informational dissolution caused by robotic or cyber warfare is not a beneficial outcome and we can measure that by the extent to which the information lost is difficult to retrieve or replace.

Finally, there is one more term that needs to be clarified before we can go on. Here we will use the term "virtue" to refer to the proper reasoning, programming, or habits of artificial and/or natural moral agents which are needed to ensure that one's actions bring about beneficial conduct. This should allow us to build an argument that in some cases a virtuous artificial agent could use deception to bring about a beneficial situation measured in terms of avoiding informational dissolution. While this notion of virtue is not precisely the same as is used in ancient or modern virtue ethics, it is still a position that is defaceable and will be useful for building the following arguments.

12.3 Informational Warfare and the Commitment to Just War

No form of warfare is ever fully just or ethical. This is due to its destructive nature; warfare always creates immediate maleficent outcomes. As the famous American civil war general Tecumseh Sherman observed, "War is all Hell."[2] At best it can only serve as a way to assure that one's enemies are dealt a greater share of that hell than they can deal to you. The destruction of warfare might also be mitigated if the war is just as it is claimed in just war theory that if the reason for the war is just and the war is fought justly and ethically, then the short term evil of the destruction and violence of war can lead to long term good in the form of a stronger and lasting peace.

If we follow this reasoning, then we must conclude that although informational warfare will always contribute to short-term maleficent outcomes in the form of irretrievable loss of life, property and information, but if these war fighting tools are used in the propagation of just war, then it might lead to long term good. This leads us to our first claim; Informational warfare must be committed to the propagation of only just war.

Technologies embody the moral commitments of their makers and users. This means that the design of informational warfare technologies can lead to systems that either enhance our commitment to just war or degrade it. The modern battlefield has evolved into a place where a great deal of information is available to the war fighter, which is good only if that information can be quickly processed and the useful and accurate information sifted from the false and useless. Acting on poor information can lead to unintended damage and casualties. Making quick and accurate decisions that lead to ethical outcomes is a taxing activity that can quickly overwhelm the cognitive capacity of unaided human agents. The job of informational warfare is to assist the war fighter in making good decisions. But it is increasingly the case that informational warfare must be more that simply data acquisition and management tools, due to the pressures to make these decisions in a faster and more efficient manner, it is inevitable that more and more of the processing and synthesis of information as well as decision making based on this information be done by the informational system itself (Singer 2009).

While the situation on the modern battlefield may demand these capabilities from our informational warfare systems, giving them this capability is much easier said than done. It is not my purpose her to outline the many obstacles to the development of these systems. Instead I wish to grant that these problems are only technical

[2] More specifically he is quoted as saying, "There is many a boy here today who looks on war as all glory, but, boys, it is all hell," at a speech given April 11, 1880 in Columbus Ohio.

issues that will be solved sooner or latter. What I want to explore here is the question of which virtues do we need to program into artificial moral agents as they become more autonomous in order to maintain our commitment to just war.

12.4 The Virtues of Informational Systems

When we talk of virtue it is easy to anthropomorphize our machines and miss apply the concept to artificial agents. Ethical systems based on the concept of virtue are all designed with the basic assumption that we are only dealing with other human moral agents. While virtue ethics is a powerful system for understanding and refining human ethical judgment, it must be adapted for use by artificial agents unless and until those agents achieve human level intelligence and become interested in human style eudemonia.

This can be illustrated by looking at Aristotle's famous illustration of virtue, courage. He argues that true courage exists in a mean position between cowardice and foolhardiness. Courage is the willingness to risk harm in the pursuit of protecting other moral agents or important ideals. The exemplar of this would be the virtuous soldier who takes risks to protect other human agents and justice, but does not simply throw his life away in a pointless gesture of bravado. What makes the behavior so exceptional and worthy of praise is that the virtuous soldier may lose her or his life in the process, so they are risking literally everything for altruistic reasons. None of this makes any sense when applied to an artificial agent such as a military robot or cyber warfare system. How can these systems display this kind of courage? They risk very little, they have no sense of existence nor do they have their own goals or desires. More importantly, they do not have beliefs about their own goals and desires which they can modify to become more virtuous in the classical sense. Thus their actions are not entirely their own and cannot be said to be motivated by anything like human courage.

Even if the same action they commit would be considered courageous if done by a human agent. Human medics and corpsmen are noted for their many acts of courage through the centuries saving wounded warfighters often while under enemy fire. Now imagine a cleverly designed and programed robot that rescues a wounded human warfighter under similar enemy fire. Would that machine be worthy of the same kind of commendation we might give a human medic or corpsman? The question here is much more difficult to answer. I have argued that "Robots are moral agents when there is a reasonable level of abstraction under which we must grant that the machine has autonomous intentions and responsibilities" (Sullins 2011b). So we might grant the machine moral agency depending on the autonomy, intentionality and responsibility of the machine in question, but it would take quite a lot of these three requirements before we might be tempted to claim that the machine was exhibiting excellence in the virtue of courage.

Of course this all changes if these systems develop, or are given the conscious understanding of their own existence and develop unique personalities that can be risked. Then these systems might also develop courage. But here we are talking about far future systems and are losing our focus on existent and near future technologies. Instead we must look at the kinds of virtues appropriate for informational systems. While these virtues are not necessarily sufficient for human moral agents, they are actually of some value when we are dealing with the restricted or even nonexistent self-awareness of artificial agents as they exist today.

Even though the long list of human virtues are barely applicable to artificial agents causing them to be seemingly impoverished moral agents, there is a potential benefit that can be leveraged. Human moral agents have many conflicting drives and desires that can complicate their ability to act entirely virtuous in any given situation. Artificial moral agents, at least the simple ones we can imagine in the near future, have a much more restricted list of potential virtues and therefore the complex internal moral conundrums should be rarer for them.

It might seem that the argument so far has concluded that traditional virtue ethics might not have much to add to our discussion, but that is only true if we are fixated on human level virtue. Instead we should shift our focus to virtues that are appropriate for artificial agents designed for informational warfare.

The virtues we are about to discuss are inspired by the "CIA" security triad that has been in use by the computer security community for some time now. The acronym "CIA" refers to: Confidentiality, Integrity, and Availability. These represent the desirable qualities that should be expressed by security systems. Confidentiality is used to insure that only authorized individuals have access to stored information. Integrity represents the ability of a security system to keep tabs on who and how any data is modified. Availability is the system's ability to have the data ready and accessible for legitimate users. This is a very sensible list but there have been many alterations to these basic concepts over the years by various interested parties. For instance the Organization for Economic Co-operation and Development (OECD) in the *OECD Guidelines for the Security of Information Systems and Networks* lists to nine separate principles for security professionals: Awareness of the need for security, Responsibility for secure information, Response to issues in a timely manner, Ethics and respect towards users, Democracy should be upheld, Risk assessment must be through, Security design and implementation in all systems and networks, Security management should reflect the above values and, Reassessment of these systems must be regular (OECD 2002). The security guru Bruce Schneier suggests this list: Privacy, Multilevel Security or secrets within secrets, Anonymity (personal and political), Commercial Anonymity, Medical anonymity, Authentication, Integrity, Audit, and Proactive solutions to threats (Schneier 2000). Taken together we can see some overlap in these lists of principles but some seem to refer to the human operators and users of the systems and some obviously refer only to the systems themselves. Next I would like to disentangle these principles with an eye towards application in informational warfare systems themselves.

It is fair to ask here why insist on using the term virtue when security professionals obviously prefer to speak of principles or rules? The main benefit of working with virtues is that they are understood to be the mean between two extremes. A virtue is a nuanced approach to moral reasoning rather than an all or nothing step function. We will see how that works out below.

Informational warfare systems need three foundational virtues; security, integrity and accessibility. Informational security is achieved when the system is able to balance the needs of integrity and accessibility demanded by systems and users at differing security levels. Simply put, data stored at a high security levels must be kept free from modification and deletion by lower security users or outside intruders. Integrity is achieved by balancing the needs of accessibility and security of data use by users of various security levels. This means that data use by low security level users or systems must not be allowed to be contaminated high security level data, though the system must be able to profit from low level information that can be verified, e.g. information obtained from an informant of some sort. Accessibility is obtained when the system correctly balances the needs of data use for all levels of systems and users while maintaining security and integrity. This requires that low security level systems and users have access to the information that they are warranted and that all of their data must be made available to higher security level systems and users if needed and where it is appropriate. In addition to this we can only claim a system is accessible when the system or user is able to access information needed precisely when needed it is needed.

These virtues having been abstracted from the civilian security profession have some interesting ethical commitments that may need modification for use in informational warfare situations. This is due to the fact that these systems must maintain security, integrity and accessibility while at the same time working to deny these very same abilities to enemy informational systems. In the civilian setting we can see from the lists of principles above that ethical security professionals have a strong desire to insure that there systems deal honestly with their legitimate users. For instance, they are not designed to give false information to certain users, but rather to simply deny access to protected information.[3] If a user has the proper security level then the system will become fully open and trustworthy. These systems are designed to be trustworthy and honest. Fine virtues indeed but if informational warfare systems adopt only these virtues, than they may be vulnerable. As we found above, deceit and the understanding that other users and systems might be potentially being deceitful to them is a necessary capability of informational warfare systems.

12.5 Robots, Informational Systems and Deceit

Research in building informational systems that intentionally deceive humans or other systems is only just beginning. Of course many forms of spy and malware work by causing the system they infect to think of them as just another benign sys-

[3] Note that as we discussed earlier in the paper, omission can be used to mislead but I do not think that security professionals are necessarily trying to fool their users in this way.

tem with all the proper security clearances. But this is a very minor form of deceit, just a kind of disguise or camouflage.

Ronald Arkin has begun to be experiment in adding deception to the capabilities of robotic weapons systems which received a good deal of media attention. In fact it received so much media attention that Arkin released a statement on the web stating some of his views on the ethical questions raised by this kind of research (Arkin 2011a).

What Arkin and his colleague in this research Dr. Alan Wagner achieved was to program their small autonomous mobile robots in such a way that they each had the ability to develop a model of what the other machine might be "thinking" was true about the toy world they were operating in and then use that information to deceive the other in a simple game of hide and seek (Wagner and Arkin 2009, 2011). These machines had the ability to do things like construct a false "trail" that the other robot would misread to look for the robot in the wrong hiding place (Wagner and Arkin 2011).

> This involves the use of partner modeling or a simplistic view (currently) of theory of mind to enable the robot to (1) assess a situation; (2) recognize whether conflict and dependence exist in that situation between deceiver and mark, which is an indicator of the value of deception; (3) probe the partner (mark) to develop an understanding of their potential actions and perceptions; and (4) then choose an action which induces an incorrect outcome assessment in the partner. (Arkin 2011a)

Perhaps one might want to quibble with Wagner and Arkin as the exact capabilities of the deceitful robots they built. It is obvious that the machines in question are only capable of deceiving one another and would not be very good at a game of hide and seek played against a human or an animal. But in an informational warfare scenario, often the target will be other computational systems so this research shows that deception of this sort is possible.

Arkin comes to some of the same conclusions seen in this paper above regarding the ethical justification for deceit in artificial systems; he agrees that there is no deontological justification but that it might be arguable on consequentialist grounds (ibid.). He does conclude that:

> The point of this paper is not to argue that robotic deception is ethically justifiable or not, but rather to help generate discussion on the subject, and consider its ramifications. As of now there are absolutely no guidelines for researchers in this space, and it indeed may be the case that some should be created or imposed, either from within the robotics community or from external forces. But the time is coming, if left unchecked, you may not be able to believe or trust your own intelligent devices. Is that what we want?. (ibid.)

Another interesting experiment using a simple autonomous robot that served as a referee in a game. The machine used the occasional strategic lie to keep the players interested and the game going. In this experiment it was really the participant's reactions that were being measured and the experimenters reported that:

> Results include the finding that participants were more accepting of lying by our robot than for robots in general. Some participants found the balancing strategy favorable after being debriefed, while others showed less interest due to a perceived level of unfairness. (Vázquez et al. 2011)

Sharkey and Sharkey (2011b) in an article for the IEEE Robots and Automation Magazine, describe how some forms of deception might be useful in the deployment of carebots for the elderly. We must note that this endorsement is only for certain situations that can clearly benefit the patient.

Peter A. Hancock et al. (2011), of the US Army Research Labs have done a nice literature review of the factors that are effecting users trust of robotic systems in military contexts. The findings of use to us here are that military robots form an integral part of human machine teams and are used to help mitigate the cognitive overload of warfighters attempting to determine accurate situational awareness in combat situations. They also find that trust is a complex human psychological state and that humans in these human-machine teams can place both too much and too little trust in their robotic assets. They also have determined that human, environmental and robot characteristics all impact the level of trust placed in the robot and negative trust can be mitigated by proper training and design (Hancock et al. 2011).

From these initial results it would seem that except for the Army Research Labs report, there is some hesitant support for allowing informational warfare systems to be engaged in some forms of deceit.

In order for that deceit to be ethical, it must be done in such a way that the resulting situation is more beneficial than would obtain had the deceit not occurred. In this context that would require at the minimum that the deceit results in a situation that advances the dictates of the rules of engagement, laws of war and principles of just war that are attendant to the conflict at hand.

As informational warfare systems become more autonomous, they must then be designed with a commitment to the foundational virtues of security, integrity and accessibility. Strategic deceit does not run counter to these virtues but in fact can help maintain them in certain situations.

The main problem we have to worry about here is that building deceit into our systems will violate the cherished notion that computers never lie. Trust and robotics has a troubled relationship (Coeckelbergh 2012). We can see from the Army Research Labs report that there are occasions already where humans working in close partnership with machines on the battlefield distrust the information they are receiving from them. If the machine was known to have the ability to deceive, then this might exacerbate the situation and make the partnership unworkable. For this reason it would be best to design the machines to error on the side of disclosure to legitimate users and only use deceit in the face of enemy threats or in actions to defeat enemy informational warfare systems.

12.6 Conclusions

This paper has shown that ethics can be profitably seen as a kind of technological undertaking designed to test and validate social values. Here we have taken on the task of validating our intuitions on the use of deceit by informational warfare systems. We found that in certain situations (but not all) deceit may be the more ethical choice in bringing about situations that fulfill our commitment to just war and the rules and laws of war. To achieve this goal we found that informational warfare systems must maintain commitments to the foundational virtues required of an informational agent; security, integrity, and accessibility. These virtues would be insufficient for a human agent but are adequate for the limited artificial agents under discussion here. Systems built with these values in mind will have the ability to retain the trust of all appropriate users who interact with the system. We also found that the virtues of an informational agent are very different from those of a human agent. A virtuous informational agent that is balancing the needs for security, integrity and accessibility needs to reveal its intentions only to its legitimate users while keeping certain bits of data confidential from low security level users, and resist the attempts of intruders into the system and other low security agents whom might wish to change its programming or stored data.

Critics of this position have been worried that any system built with these capabilities might move beyond the control of the human agents deploying it or might even be cynically used by humans in a way to deny responsibility for any harm committed by the informational warfare system. One should not deeply worry about the responsibility gap for the commitment of war crimes as argued by Robert Sparrow (see, Sparrow 2007). It would be unrealistic to let the owners and operators of some informational warfare machine off the hook due to the autonomy of the systems deployed. This issue is addressed nicely by a workgroup from the US National Science Foundation and their findings are summed up in a document informally known as "The Rules" and rule 1 clearly states that: The people who design, develop, or deploy a computing artifact are morally responsible for that artifact, and for the foreseeable effects of that artifact.

Finally, while the virtuous lie is an old idea first formally argued for by Plato in The Republic. It has always been a very fraught move morally. One always has to ask if they are telling the lie for the good of the other or simply as an expedient for themselves. It is very easy to fool oneself into thinking their lie is virtuous while those told to them are villainous. But an artificial agent should be actually more successful at navigating the moral paradox of the virtuous lie, given that it has no real stake in the game, no actual wants needs or desires. It is more likely to avoid motivational conflicts common in human agents.

Bibliography

Anderson, M., and S. L. Anderson, eds. 2011. *Machine ethics*. Cambridge: Cambridge University Press.

Arkin, R. 2009a. *Governing lethal behavior in autonomous robots*. New York: Chapman and Hall Imprint, Taylor and Francis Group.

Arkin, R. C. 2009b. Ethical robots in warfare. *IEEE Technology and Society Magazine* 28(1):30–33 (Spring 2009).

Arkin, R. C. 2010. The case for ethical autonomy in unmanned systems. *Journal of Military Ethics* 9(4):332–341.

Arkin, R. C. 2011a. The ethics of robotic deception. http://www.cc.gatech.edu/ai/robot-lab/online-publications/deception-final.pdf. Accessed July 2012.

Arkin, R. C. 2011b. Viewpoint: Military robotics and the robotics community's responsibility. *Industrial Robotics* 38(5).

Arkin, R. C., P. Ulam, and A. R. Wagner. 2012 Mar. Moral decision-making in autonomous systems: Enforcement, moral emotions, dignity, trust and deception. *Proceedings of the IEEE* 100(3):571–589.

Arquilla, J. 2010. Conflict, security and computer ethics in Floridi 2010.

Aycock, J., and J. Sullins. 2010. Ethical proactive threat research. Workshop on Ethics in Computer Security Research (LNCS 6054), pp 231–239. New York: Springer.

Bunge, M. 1977. Towards a technoethics. The Monist, 60, 96–107.

Cisco Systems Inc. 2011. Cisco 2011 Annual Security Report: Highlighting global security threats and trends. San Jose: Cisco Systems Inc.

Coeckelbergh, M. 2012. Can we trust robots? *Ethics and Information Technology* 14:53–60.

Crnkovic, G. D., Çürüklü, B. 2012. Robots: Ethical by design. *Ethics and Information Technology* 14:61–71.

Denning, D. 2008. The ethics of cyber conflict. In *The handbook of information and computer ethics*. 1st ed, ed. K. E. Himma, and H. T. Tavanni. Wiley-Interscience.

Dietrich, E. 2011a. There is no progress in philosophy. *Essays in Philosophy* 12 (2).

Dietrich, E. 2011b. Homo Sapiens 2.0 why we should build the better robots of our nature. In *Machine ethics,* ed. M. Anderson, and S. Anderson. Cambridge: Cambridge University Press.

Floridi, L., ed. 2010. *The Cambridge handbook of information and computer ethics*. Cambridge: Cambridge University Press.

Floridi, L. 2011. *The philosophy of information*. Oxford: Oxford University Press.

Gouinlock, J. 1994. *The moral writings of John Dewey*. Buffalo: Prometheus.

Hancock, P. A., D. R. Billings, K. E. Oleson, J. Y. C. Chen, E. De Visser, R. Parasuraman. 2011. A meta-analysis of factors influencing the development of Human-Robot Trust, Army Research Laboratory, ARL-TR-5857, December 2011. http://www.dtic.mil/cgi-bin/GetTRDoc?AD=ADA556734. Accessed July 2012.

Hickman, L. A. 1990. *John Dewey's pragmatic technology*. Bloomington: Indiana University Press.

Himma, K. E., ed. 2007. *Internet security, hacking, counterhacking, and society*. Sudbury Massachusetts: Jones and Bartlett.

Himma, K. E., and H. T. Tavanni, eds. 2008. *The handbook of information and computer ethics*. Wiley-Interscience. 1st edition.

Kaspersky Lab. 2011. *Cyberthreat forecast for 2012*. Moscow: Kaspersky Lab ZAO (available online).

Lin, P., G. Bekey, and K. Abney. 2008. *Autonomous military robotics: Risk, ethics, and design*. Washington, DC: US Department of the Navy, Office of Naval Research (available online).

Lin, P., K. Abney, and G. Bekey. 2011. *Robot ethics: The ethical and social implications of robotics*. Cambridge: MIT Press.

Lovely, E. 2010 Mar 5. Cyberattacks explode in Congress. *Politico* (available online).

Marchant, G., B. Allenby, R. Arkin, E. Barrett, J. Borenstein, L. Gaudet, O. Kittrie, P. Lin, G. Lucas, R. O'Meara, and J. Silberman. 2011. International governance of autonomous military robots. *Columbia Science and Technology Law Review* XII:272–315.

Miller, K. W. 2011. Moral responsibility for computing artifacts: The rules. *IT Professional* 13(3):57–59.

OECD. 2002. OECD Guidelines for the Security of Information Systems and Networks: Towards a Culture of Security, Organisation for Economic Co-operation and Development, Organisation De Coopération Et De Développement Économiques. http://www.oecd.org/dataoecd/16/22/15582260.pdf. Accessed July 2012.

Ramaswamy, S., and H. Joshi. 2009. *Automation and Ethics. Springer Handbook of Automation*. Springer.

Shannon, C. E., and W. Weaver. 1949. *The mathematical theory of communication.* University of Illinois Press.

Sharkey, N. E. (2007 Nov). Automated killers and the computing profession. *IEEE Compute* 40 (11):106–108

Sharkey, N. E. (2008a). Grounds for discrimination: Autonomous robot weapons. *RUSI Defence Systems* 11 (2):86–89.

Sharkey, N. E. (2008b). Cassandra or the false prophet of doom: AI robots and war. *IEEE Intelligent Systems* 23 (4):14–17 (JulyAugust Issue).

Sharkey, N. E. (2009a). Death strikes from the sky: The calculus of proportionality. *IEEE Science and Society* 28:16–19.

Sharkey, N. E. (2009b). Weapons of indiscriminate lethality. *FIfF Kommunikation* 1:26–28

Sharkey, N. E. (2010). Saying "No!" to lethal autonomous targeting. *Journal of Military Ethics* 9 (4):299–313.

Sharkey, N. E. (2011a). Killing made easy: From joystics to politics. In *Robot ethics: The ethical and social implications of robotics,* eds. Lin Patrick, George Bekey, and Keith Abney. Cambridge: MIT Press

Sharkey, A. J. C., and N. E. Sharkey. 2011b. Anthropomorphism and deception in robot care and companionship. *IEEE RAM* 18 (1):32–38.

Schneier, B. 2000. *Secrets and lies: Digital security in a networked world.* New York: Weiley.

Singer, P. W. 2009 *Wired for war: The robotics revolution and conflict in the 21st century.* New York: Penguin.

Sparrow, R. 2007. Killer robots. *Journal of Applied Philosophy* 24 (1):62–77.

Sullins, J. P. 2009. Telerobotic weapons systems and the ethical conduct of war. *APA Newsletter on Philosophy and Computers* 8 (2):21 (P. Boltuc, ed.).

Sullins, J. P. 2011a. Robotics: War and peace. *Philosophy and Technology* 24(3)September.

Sullins, J. P., ed. 2011b. When is a robot a moral agent? In Machine ethics, ed. M. Anderson, and S. L. Anderson.

Tavani, H. T. 2007. The conceptual and moral landscape of computer security. In *Internet security, hacking, counterhacking, and society,* ed. K. E. Himma, 29–45. Sudbury Massachusetts: Jones and Bartlett.

Vázquez, M., A. May, A. Steinfeld. (2011). ShakeTime! A deceptive robot referee. Copyright is held by the author/owner(s), HRI'11, March 6–9, 2011, Lausanne, Switzerland, ACM 978–1-4503-0561-7/11/03. http://delivery.acm.org/10.1145/1960000/1957803/p403-vazquez.pdf?ip=130.157.156.155&acc=ACTIVE%20SERVICE & CFID=97012119 & CFTOKEN=57280102&__acm__=1342826463_29deff0d3692bce36f2a3436192d93e8. Accessed July 2012.

Wagner, A., and R. C. Arkin. (2009). Robot deception: Recognizing when a robot should deceive. *Proc. IEEE International Symposium on Computational Intelligence in Robotics and Automation (CIRA-09),* Daejeon, KR.

Wagner, A. R., and R. C. Arkin. 2011. Acting deceptively: Providing robots with the capacity for deception. *International Journal of Social Robotics* 3 (1):5–26.

Wallach, W., and C. Allen. 2010. *Moral machines: Teaching robots right from wrong.* Oxford: Oxford University Press.

ERRATUM

The Ethics of Informational Warfare

Mariarosaria Taddeo[1] · Luciano Floridi[2]

[1]Politics and International Studies
Research Fellow in Cyber Security and
Ethics, PAIS
University of Warwick
Coventry
United Kingdom

[2]Professor of Philosophy and
Ethics of Information
Oxford Internet Institute
University of Oxford
Oxford
United Kingdom

M. Taddeo, L. Floridi (eds.), *The Ethics of Information Warfare,* Law,
Governance and Technology Series 14, DOI 10.1007/978-3-319-04135-3,

---

DOI 10.1007/978-3-319-04135-3_13

The Publisher regrets that in the online version of this book the introduction to volume 'The Ethics of InformationWarfare—An Overview' did not reflect the book editors' name: Mariarosaria Taddeo and Luciano Floridi.

The online version of the original book can be found at
http://dx.doi.org/10.1007/978-3-319-04135-3

Law, Governance and Technology Series 14, DOI 10.1007/978-3-319-04135-3_13,

Afterword

Robots and Other Cognitive Systems: Challenges and European Responses

Robots have come a long way since the Czech writer Karel Čapek first used this term, some 90 years ago, to denote rather frightening creatures—not unlike Golems or Frankenstein's monster, yet workers all the same. Today, more than ever, robots continue to fascinate: they take over activities which humans find too dangerous or impossible. For example, the recent use of robots at Japan's Fukushima nuclear power plant or in the recovery of the flight recorder of Air France's Rio de Janeiro—Paris flight which went down in deep seas in 2009. They go to war and deactivate mines. And they increasingly come into our homes as children's toys, almost like family pets!

In just a few years, technological progress in this area has been tremendous and Europe is one of the leaders in this research and industrial application. Yet, this is just the beginning of the robot history as many challenges remain to be addressed. Refining and improving the mechanics of robots and their sensorial capacities (including ones that living organisms do not possess) has always been of major concern for engineers. Reducing the amount of human intervention in the operation of these machines has been another persistent trend, leading for instance to numerically controlled machine tools. Ultimately, however, this means more than merely automating the completion of a task according to some preset rules. It means that, within certain limits, machines ought to be able to take "decisions" autonomously and independent of external (e.g., remote) control on how to proceed with a given task should new conditions arise unexpectedly. This could be in the form of a roving robot that is supposed to retrieve some object from a distant place but on its way encounters an unexpected obstacle.

The ease of use, safety, and partial autonomy are essential if robotic devices are to leave the shop floor and strictly controlled environments and become truly

---

The content of this chapter has been previously published. It first appeared in September 2011 in *Philosophy & Technology* 24(3), 355–357.

L. Floridi, M. Taddeo (eds.), *The Ethics of Information Warfare*,
Law, Governance and Technology Series 14, DOI 10.1007/978-3-319-04135-3,

useful and helpful for people, including those with special needs. This could include steering a wheelchair, driving a car, guiding a blind person, performing precision surgery, operating a leg amputee's prosthesis, or many of our everyday chores.

None of these machines are expected to solve chess conundrums or any other classical artificial intelligence problem. But they should have their wits about them, if for instance, they might need to recognise a certain object viewed from a different angle or under different lighting conditions. Other systems will need to understand their users' intentions and what they are saying in plain natural language. All of them would have to understand to a greater or lesser degree the aspects and features of their environment. We may for instance want robots to "know" or be able to "learn" what they can do with certain objects of our world: what the handle of a mug is for or a dishwasher or the curb of the pavement along a busy street…

Machines and systems which are cognitive are still far from being as intelligent or conscious as humans or animals of what they are doing. Engineers have a lot to learn to catch up with solutions that natural evolution has developed over billions of years.

Considerable research effort taking new, multidisciplinary approaches are needed to significantly advance the engineering of the machines and systems described above. From the very start, the European Union's Framework Programmes for Research and Technical Development has acknowledged the potential of robotics and cognitive systems research for increasing the productivity of human labour and creating new useful products and services.

Cognitive systems were one of the key challenges in the Information Society Technologies chapter of the ixth Framework Programme which ran from 2002–2006. In the current EU research programme (FP-7), the scope has been broadened to cognitive systems and robotics and given even more weight in the Information and Communication Technologies (ICT) programme.[1] More lines of relevant basic research have been opened up in the Future and Emerging Technologies (FET) part of this programme.[2]

Currently, about 100 research grants, falling within the remit of the FP7-ICT Cognitive Systems and Robotics challenge, have been or will shortly be awarded to the consortia of European researchers. Funded projects address general issues related to endowing artificial systems with cognitive capabilities and issues specifically related to the design of all kinds of robots.

RoboCom, one of those robotics projects, is amongst the six finalists competing for the chance to become a "FET" flagship.[3] It proposes an ecology of sentient machines that will validate our understanding of the general design principles underlying biological bodies and brains, establishing positive feedback between science and engineering.

---

[1] http://cordis.europa.eu/fp7/ict/.

[2] http://cordis.europa.eu/fp7/ict/programme/fet_en.html.

[3] http://europa.eu/rapid/pressReleasesAction.do?reference=MEMO/11/270&format=HTML&aged=0&language=EN&guiLanguage=en.

These projects add to Europe's knowledge and prowess in building robotic systems which are ever safer, more robust, efficient, easy to use, and—where needed—more autonomous. Projects also develop the features needed so these systems could be used in a wide range of scenarios: industrial/service/medical robotics, land, submarine or spatial exploration, logistics, maintenance and repair, search and rescue, environmental monitoring and control, physical and cognitive assistance to disabled and elderly people, and many more.

But some questions remain. We cannot and must not curb scientific curiosity but we should ask: are there general principles that might guide public funding of research and the use of its results beyond innovation and competitiveness?

Take for instance the concept of an autonomous machine. This could be a self-controlling road vehicle, which may become a reality sooner rather than later given the current speed of technological advancement. There are also various examples of military autonomous vehicles operating on land, at sea or in the air. Who is responsible for their actions? Who is liable in case of damage? Can it be considered that such machines operate on their own accord?

The answer is a firm "no". Machines are designed, built and programmed so that they can render services. They are always owned and controlled by people. Machines—no matter how sophisticated—are as "ethical" as the people who design, build, programme and use them.

We humans, jointly and individually, have to take full responsibility for what we are doing, good or bad, constructive or destructive, through our own inventions and creations, to each other and our world at large.

Bertolt Brecht, in "The Life of Galilei", had the great scientist say: "I maintain that the only goal of science is to alleviate the drudgery of human life." Sound advice indeed! We will continue to fund research whose results help create better living conditions for everyone on this planet and research that helps us to better understand ourselves and the world we live in. Both go hand in hand—and robots should take their fair share in this ICT landscape.

European Commission's Vice President
for the Digital Agenda,
200 Rue de la Loi, Brussels, Belgium
neelie.kroes@ec.europa.eu

Neelie Kroes

About the Authors

Keith Abney is a senior philosophy lecturer and research fellow at the Ethics + Emerging Sciences Group at Cal Poly. He has published numerous articles and worked on several projects related to military-technology ethics and cyberpolicy, including co-authorship on an ONR-funded ethics report on autonomous military robots and a Greenwall Foundation-funded ethics report on human enhancement technologies in the military, as well as co-authoring the book *Robot Ethics*. He also is active in the Consortium for Emerging Technologies, Military Operations, and National Security (CETMONS). Currently, he is a senior investigator on a NSF-funded project on the ethics of cyberwarfare. He earned his B.A. in philosophy from Emory University and M.A. (A.B.D.) from Univ. of Notre Dame.

Fritz Allhoff is Associate Professor in the Department of Philosophy at Western Michigan University and Senior Research Fellow at the Centre for Applied Philosophy and Public Ethics, Charles Sturt University (Australia). He has held visiting appointments at the University of Oxford, the University of Michigan, and the University of Pittsburgh. His research principally involves the ethics of war and the ethics of technology, including military technologies. The author or editor of nearly thirty books, two of his most recent are *Terrorism, Ticking Time-Bombs, and Torture* (University of Chicago Press, 2012) and *The Routledge Handbook of Ethics and War* (2013). Dr. Allhoff was a founding member of the International Intelligence Ethics Association, serves on the advisory board for the International Committee of Military Medicine (Switzerland), and is active in the Consortium for Emerging Technologies, Military Operations, and National Security (CETMONS). Currently, he is a principal investigator on a NSF-funded project on the ethics of cyberwarfare.

Deen Chatterjee is Senior Advisor and Professorial Fellow in the S. J. Quinney College of Law at the University of Utah and the editor-in-chief of the two-volume Encyclopedia of Global Justice (2011) and the series editor of Studies in Global Justice. His areas of specialization are justice and global initiative, ethics of war and peace, and philosophy of religion and culture.

Dorothy E. Denning is Distinguished Professor of Defense Analysis and a member of the Cyber Academic Group at the Naval Postgraduate School. Her

L. Floridi, M. Taddeo (eds.), *The Ethics of Information Warfare*, Law, Governance and Technology Series 14, DOI 10.1007/978-3-319-04135-3,

teaching, research, and publications have been mainly in the area of cyber security and cyber conflict, and include contributions in intrusion detection, information flow security, database security, cryptography, hacktivism, cyber terrorism, and cyber ethics.

Randall R. Dipert is Charles S. Peirce Professor of Philosophy at the University at Buffalo. In 2011–2012 he was a Fellow at the Stockdale Center for Ethical Leadership at the U.S. Naval Academy for research on the ethics of cyberwarfare. From 1995 to 2000 he taught at the United States Military Academy at West Point, NY and he has worked on applied ontology for the U.S. Army. He has published a book on artifacts and action theory, and has co-authored a book on logic. He has published dozens of articles on Peirce, logic, artifacts, mathematics, the philosophy of mind, artificial intelligence, and ontology. He has published papers and given presentations on the morality of preventive war and the application of game theory to the morality of war. He authored one of the first papers on the subject "The Ethics of Cyberwarfare" in the J. of Military Ethics (December 2010).

Luciano Floridi is Professor of Philosophy and Ethics of Information at the University of Oxford, Senior Research Fellow at the Oxford Internet Institute, and Fellow of St Cross College, Oxford. Among his recognitions, he was the UNESCO Chair in Information and Computer Ethics, Gauss Professor of the Academy of Sciences in Göttingen, and is recipient of the APA's Barwise Prize, the IACAP's Covey Award, and the INSEIT's Weizenbaum Award. He is an AISB and BCS Fellow, and Editor in Chief of Philosophy & Technology and of the Synthese Library. He was Chairman of EU Commission's "Onlife Initiative". His most recent books are: The Fourth Revolution—How the infosphere is reshaping human reality (Oxford University Press, 2014), The Ethics of Information (OUP, 2013), The Philosophy of Information (OUP, 2011), The Cambridge Handbook of Information and Computer Ethics (editor, CUP, 2010), and Information: A Very Short Introduction (OUP, 2010).

Don Howard is Professor of Philosophy and Director of the Reilly Center for Science, Technology, and Values at the University of Notre Dame. A historian and philosopher of science who has written extensively on Einstein, Bohr, and the history and foundations of twentieth-century physics as well as the history of the philosophy of science, Howard also works on a wide range of science and technology ethics questions, including the ethics of emerging weapons technologies and the ethics of autonomous robotic systems. Among his recent publications is the 2008 volume, *The Challenge of the Social and the Pressure of Practice: Science and Values Revisited* (co-edited with Martin Carrier and Janet Kourany).

Neelie Kroes is the vice president of the European Commission and European Digital Agenda Commissioner. From 2004 to 2009 she worked as advisor for the Nelson Mandela Children's Fund and World Cancer Research Fund. She was also appointed President of Nyenrode University, and served on various company boards, including Lucent Technologies, Volvo, P&O Nedlloyd. From 1982–1989 she served as Minister for Transport, Public Works and Telecommunication in the Netherlands.

Steven P. Lee is professor of Philosophy at Hobart and William Smith Colleges. His research interests include issues in social, moral, and political philosophy, especially on matters of morality and war. He has done extensive work on the ethical challenges posed by nuclear weapons. He is the author of *Nuclear Weapons, Nuclear States, and Terrorism* (with Cornwall-on-Hudson), Sloan Publishing, 2007; *Morality, Prudence, and Nuclear Weapons*, Cambridge University Press, 1993–1996. He has edited *Intervention, Terrorism, and Torture: Contemporary Challenges to Just War Theory*, Springer, 2007; *Ethics and War*, Cambridge University Press, forthcoming; and *Ethics and Weapons of Mass Destruction* (with Sohail Hashmi), Cambridge University Press, 2004.

Patrick Lin is the director of the Ethics + Emerging Sciences Group at California Polytechnic State University, San Luis Obispo, where he is an associate philosophy professor. He is also a visiting associate professor at Stanford's School of Engineering; an affiliate scholar at Stanford Law School's Center for Internet and Society; and an adjunct senior research fellow at CAPPE. Previously, he was the interim executive director of IACAP, an ethics fellow at the US Naval Academy, and post-doctoral associate at Dartmouth College—all with a focus on technology ethics. His current work on cyberwarfare includes: invited briefings to organizations such as UNESCO, National Research Council, DARPA, and others, as well as several published papers. Dr. Lin is well published in the technology ethics, including co-editing or co-authoring four books. He was also the lead author on an ONR-funded ethics report on autonomous military robots and a NSF-funded ethics report on human enhancement technologies. Currently, he is a principal investigator on a NSF-funded project on the ethics of cyberwarfare. He earned his B.A. from UC Berkeley and Ph.D. from UC Santa Barbara.

George R. Lucas, Jr., is Class of 1984 Distinguished Chair in Ethics at the Stockdale Center for Ethics, U.S. Naval Academy (retired), and Professor of Ethics & Public Policy at the Naval Postgraduate School (Monterey, CA), He is author of *ANTHROPOLOGISTS IN ARMS: The Ethics of Military Anthropology* (AltaMira Press, 2009), editor and contributor to *NEW WARRIORS/NEW WEAPONS: Ethics and Emerging Military Technologies*, a special issue of the *JOURNAL OF MILITARY ETHICS* (9, no. 4: December 2010), and editor of the Ethics Section of the *HANDBOOK ON UNMANNED VEHICLES*, forthcoming from Springer Verlag (2014), as well as numerous subsequent articles on ethics and cyber conflict.

Wayne McCormack is E. W. Thode Professor of Law in the S. J. Quinney College of Law at the University of Utah. He received a B.A. from Stanford University and a J.D. from the University of Texas, where he graduated Order of the Coif and was associate editor of the Texas Law Review. After graduation he clerked for Judge Walter Ely of the U.S. Court of Appeals for the Ninth Circuit and then taught at the University of Georgia School of Law. He also served as Associate Director of the Association of American Law Schools from 1975–1978. Professor McCormack joined the faculty at the University of Utah S. J. Quinney College of Law in 1978 and served as Associate Dean for Academic Affairs from 1978–1982, 1984–1987, and 1993–1994.

Brian Orend is the Director of International Studies, and a Professor of Philosophy, at the University of Waterloo in Canada. His Ph.D. is from Columbia University in New York City. His research and speaking efforts concentrate on three areas: the ethics of war and peace (especially post-war reconstruction); human rights; and happiness. He is the author of five books, including two books used widely as required texts at colleges and universities around the world: *Human Rights: Concept and Context*, Broadview Press, 2002; and *The Morality of War*, Broadview Press, 2006. He is forthcoming with two more books: *An Introduction to International Studies*, Oxford University Press, and *Seizure the Day: Happiness in Spite of Illness*.

Thomas Simpson is University Lecturer in Philosophy and Public Policy at the Blavatnik School of Government, University of Oxford, and a Senior Research Fellow at Wadham College. He was educated at Cambridge (BA, MPhil, PhD), where he was also previously a Research Fellow at Sidney Sussex College. Between degrees he served as an officer with the Royal Marines Commandos, with tours of duty in Northern Ireland; Baghdad, Iraq; and Helmand Province, Afghanistan. His research in applied ethics has focused hitherto on the ethics of information and computing technologies, and of war.

Bradley Jay Strawser is Assistant Professor of Philosophy in the Defense Analysis Department at the U.S. Naval Postgraduate School in Monterey, California and a Research Associate with Oxford University's Institute for Ethics, Law, and Armed Conflict. His research focus is primarily ethics and political philosophy, though he has also written on metaphysics, ancient philosophy, and human rights. He recently published *Killing By Remote Control: The Ethics of an Unmanned Military* (New York: Oxford University Press, 2013), an edited volume on the many moral issues raised by drone warfare.

John P. Sullins is an associate professor of philosophy at Sonoma State University in California where he has taught since 2004. He received his PhD in 2002 from the Philosophy, Computers, and Cognitive Science program at Binghamton University in New York. His current research and publications involve the study of computer ethics, malware ethics, and the analysis of the ethical impacts of emerging technologies in both military and civilian contexts. He is the 2011 recipient of the Herbert Simon Excellence in Research award from the International Association of Computers and Philosophy.

Mariarosaria Taddeo is Fellow in Cyber Security and Ethics and the Department of Politics and International Studies at the University of Warwick and Research Associate at the Uehiro Centre for Practical Ethics, University of Oxford. Her recent work focuses mainly on the ethical analysis of cyber security practices and information conflicts, although she has worked on issue concerning Philosophy of Information, Epistemology, Philosophy of AI and Applied Ethics. Dr. Taddeo is the 2010 recipient of the *Simon Award for Outstanding Research in Computing* and Philosophy and is the President of the International Association of Computing and Philosophy.

Shannon Vallor is Associate Professor of Philosophy at Santa Clara University in California, where she teaches courses in the philosophy of science and technology, phenomenology, and engineering ethics. Her research on emerging technologies and their impact on the moral virtues has appeared in numerous peer-reviewed journals and edited collections, and she has presented her work on the ethics of military robotics and information warfare at several international conferences. She is currently writing a book on emerging technologies and moral self-cultivation, titled *21st Century Virtue*. She also serves on the Executive Board of the international *Society for Philosophy and Technology*.

